Britannia's Sceptre

Scotland and the Trident System

Brian P Jamison

Argyll
publishing

First Published 2006
Argyll Publishing
Glendaruel
Argyll PA22 3AE
Scotland
www.argyllpublishing.com

Permission to use an extract from Alasdair Gray *Why Scots Should Rule Scotland* on pages 13 and 14 is acknowledged to Canongate Books, 14 High Street, Edinburgh EH1 1TE.

British Library Cataloguing-in-Publication Data.
A catalogue record for this book is available from the British Library.

ISBN 1 902831 92 6

Printing & Binding Bell & Bain Ltd, Glasgow

This work is dedicated to the Balloch community
on the shores of Loch Lomond.
For our friends at Fairview Cottage and Haldane Terrace
it must be said that their continued generosity and graciousness
only strengthens further our bond with the place.

Resolution class Polaris SSBNs

HMS *Resolution* active from October 1967 to October 1992
HMS *Repulse* active from September 1968 to August 1996
HMS *Renown* active from November 1968 to 1995
HMS *Revenge* active from December 1969 to May 1992

Vanguard class Trident SSBNs

HMS *Vanguard* from October 1992
HMS *Victorious* from May 1994
HMS *Vigilant* from March 1996
HMS *Vengeance* from May 1999

Abbreviations and Acronyms

ABM	Anti-Ballistic Missile
AESG	Alternative Employment Study Group
ASDS	Advanced SEAL Delivery Systems
CND	Campaign for Nuclear Disarmament
DoD	Department of Defence
GLCM	Ground-Launched Cruise Missile
MIRV	Multiple Independent Re-entry Vehicle
MoD	Ministry of Defence
NATO	North Atlantic Treaty Organisation
NFLA	Nuclear Free Local Authorities
NNS	National Security Strategy
NOPD	Notice of Proposed Development
RAF	Royal Air Force
RNAD	Royal Naval Armament Depot
SCND	Scottish Campaign for Nuclear Disarmament
SDP	Social Democratic Party
SLBM	Submarine Launched Ballistic Missile
SNP	Scottish National Party
SOF	Special Operations Forces
SSBN	Surface to Surface Ballistic Nuclear
SSGN	Surface to Surface Guided Nuclear
SSN	Ship Submersible Nuclear

Contents

Preface

AT THE MOUNTBATTEN CENTRE for International Studies (University of Southampton) I am working under Professor John Simpson in an effort to produce an authoritative study of the history of development of the British nuclear weapon capability. His knowledge of such matters is nothing less than stunning and conversations with him and employees of AWE Aldermaston, the Ministry of Defence and elsewhere have provoked a fresh interest in my own research.

It has become increasingly apparent to me that Aldermaston's scientists have yet to receive the recognition they so richly deserve, but I have come to the conclusion that the same could be said for the civilian employees of Scotland's shore establishments. Rosyth Royal Dockyard in Fife performed the dirty refit work on older Polaris SSBNs; these boats tended to be the most radioactive. Faslane in Dunbartonshire served as the homeport for both the Resolution and Vanguard class SSBN; nuclear warhead storage facilities were constructed nearby at Coulport. Yet the constant threat of redundancy, in some form or another, has lingered over these facilities for some time and will continue to do so because of expanding strategic requirements and developing technologies. With a decision on Trident's replacement drawing near it has reinvigorated my initial concerns for the workers at these facilities.

Like most people I am also deeply concerned about security for the system itself. On 11 September 2001 I remember hearing what I thought at that time were Tornado long-range interceptors scrambling towards Faslane minutes after United Flight 175 from Boston crashed into the South Tower. I could be mistaken and only the MoD or the RAF can confirm this assumption. Nevertheless, over the years I have spent countless hours both in Helensburgh and around the Faslane and Coulport facilities, using that time to observe the area, interact with locals and collect my thoughts. Living in Balloch made it very convenient to visit both of the shore establishments, and over the years a friend in Blairmore kept me informed of the SSBN movements. I would notice Sea King helicopters from the Base hovering over the Vale of Leven hospital in Alexandria, and Trident's nuclear convoys rumbled past my front window on a regular basis. Sometimes this would occur in the morning, other times they would pass in the night. The signature sounds of their powerful motors are easily identified. There are indicators which make one aware of their impending journey, and it requires little effort to know when to expect these vehicles. I am a 'friendly', and, as a guest in this country, I obey completely the laws of the land. However, should it not be somewhat disconcerting for British authorities that a foreigner could become so familiar with the various activities attached to the national deterrent?

The majority of my research on Scotland's experience with Trident resides within a thesis submitted to the University of Glasgow for the degree of Doctor of Philosophy. It was Christopher Harvie who suggested that I contact Argyll Publishing, and after Derek Rodger suggested writing something about Trident for the general reader I began the production of drafts. This is the result.

A publication geared toward a general readership has its benefits because it may spark a greater sense of public awareness rather than being picked over within academic circles and left to be shelved. However, there may be questions about my intentions with this publication so I feel it necessary to address certain issues.

Having lived in Britain for seven years now, I have formed the view that a British Prime Minister needs to kick an American president in the ass from time to time. I am not a member of the disarmament movement but I recognise the importance of passionate, well-informed objection. I am not a pacifist because there are times when fighting is necessary. I feel that Britain's security is an immovable requirement and not an undemanding alternative. I believe nuclear weapons once served a purpose but now I have questions. I also think a Scottish Parliament should make known its interest both in the security of the Trident system and the creation of employment alternatives. I hope readers will come up with further questions of their own.

This book is a critical examination of Scotland's relationship with nuclear weapons. It is concerned to show how Scottish civil society reacted to Trident; why an undetermined number of Scots have chosen to either support, disregard or reject the system; how the Trident Works Programme impacted the country; where and how certain projects took place; and what was achieved, and at what costs. It also provides a view of what might replace Trident, and circumstances that could bring about the end of operations in Scotland.

Ultimately, the assignment is to open Scotland's eyes to the troubling dependence it has had on the defence industry, and how this ever-changing business, and strategic requirements, might influence the country in the future. Scotland's media, its political representation and even academia appear to have only a limited interest in these issues. Therefore, it is hoped this publication will encourage a noticeable dialogue on Scotland's future with such matters, and perhaps inspire researchers to take a greater interest in Scotland's relationship with the British military establishment and NATO.

Brian P Jamison
Southampton
November 2005

1
Wake up: time for work

> Today, our fellow citizens, our way of life, our very freedom came under attack . . . America and our friends and allies join with all those who want peace and security in the world, and we stand together to win the war against terrorism.
>
> Statement by the President in His Address to the Nation (September 11, 2001)

PILOTED BY FANATICS, the indescribable horror of watching passenger airliners accelerate and explode into skyscrapers and the Pentagon are images that will forever remain in people's minds. Perched directly beside this nightmarish scene of events, the sight of stranded and desperate people leaping to their deaths before the collapse of the World Trade Centre only bolstered the revulsion that most of the world had for that despicable act.

But what have been the repercussions of September 11? Amongst a number of developments the Taliban's grip on Afghanistan was torn away; Saddam Hussein was removed from power after being falsely (or, incorrectly) accused of WMD proliferation; the American-led war on terror created difficulties for independent humanitarian action; and the West now needs to pay particular attention to supporting Muslim countries that have turned toward democracy. However the far-reaching influence of September 11 could yet make its way into Scotland in ways many Scots might not yet appreciate fully.

Despite its proven inability to reverse persistent economic difficulties in Scotland or address past methods of IRA intimidation, questioning over the continuance of Britain's nuclear deterrent, and the procurement of its successor, will most likely be tied to inflated narratives on job creation and counter-terrorism.

Arguably, Britain's 'minimum deterrent' was slowly approaching its sell-by date after the early 1990s with the door to unilateral disarmament not yet ajar but perhaps unlocked. Having a powerful tool at their disposal central government will cite Washington's accusations of North Korea running a secret nuclear weapons programme and, in all likelihood, use attacks in London, New York and elsewhere to discourage support for disarmament, the decommissioning of Trident and any resentment towards that which is chosen to replace it. Nevertheless, given Scotland's long-established discomfort with nuclear weapons, residents north of the Border might wish to consider their country's experience with submarine platforms and the future of the programme currently in place.

For over fifty years UK nuclear weapon programmes have been shrouded in official secrecy and during the past two decades the west of Scotland has provided a convenient sanctuary for the latest British national deterrent system, Trident. This programme includes four Surface to Surface Ballistic Nuclear (SSBN) vessels with submarine-launched ballistic missiles (SLBM), and is the Royal Navy's most recent strategic and sub-strategic sea-borne nuclear deterrent. However, Scotland's longstanding relationship with submarine platforms and nuclear missile systems predates Trident.

With the arrival of the American Polaris fleet in the Holy Loch in March 1961, Scots were propelled into the mix of harsh realities that accompanied the Cold War. The entrance of the US Navy's Hunley class submarine tenders and Polaris SSBNs near Dunoon, followed by the Royal Navy's own version in the Gareloch in 1968, served as the country's formal introduction to the SSBN. These events heightened concerns with nuclear weaponry in the west of Scotland; yet apprehension was met with acquiescence. The safe operation of US and UK systems from the Strathclyde Region, coinciding with their employment potential, subdued many of the diversified concerns associated with the deterrent. Nevertheless, one might have expect to hear the phrase 'ye cannae spend a dollar when ye're deid' from a Glasgow peace activist or people who ultimately rejected central government's economic arguments for Trident.

Because Strathclyde served as Trident's operational centre, the explicit targeting of Scottish territory by Soviet military strategists during the Cold War era and the fears that accompanied the possibilities for accidental radiological contamination were of great concern. However, the vulnerability of this system to acts of terrorism might also serve to fuel ongoing trepidation. Despite rigorous safety guidelines enforced over the course of its life expectancy, Trident, and future versions of the nuclear deterrent, will undoubtedly generate controversy within Scotland so long as a system remains in place.

Taken from a publication that was an imagined transcript of a conversation between publisher and author, when asked if he was converted to Scottish Home Rule by the placing of the American and British nuclear submarine bases on the Clyde, Alasdair Gray wrote:

> Most British Labour voters did not see why the USA, when she was the only nation to possess atomic weapons, refused to sign an international agreement banning their manufacture and use. Still less did we see why Britain (which now had no empire to defend) was joining an arms race with the two biggest empires which remained: especially when Japan and West Germany were becoming the world's foremost industrial nations because they were excluded from such weapon making. We could not understand it – there seemed no explanation but human blindness. Leading politicians in those days spoke as if nuclear war was a thing civilisation could survive, while at the same time building vastly expensive bunkers for themselves and their adherents all over the country. Bunkers for Scottish administrators were built under Edinburgh Zoo and the Glasgow Burrell Collection Gallery.
>
> We were told that without nuclear weapons Britain would become second rate like Japan, West Germany and Scandinavian nations where the ordinary standard of living had risen mysteriously higher than that of Britain. Despite this

> publicity our trade unions and local Labour parties were so convinced by CND arguments that the 1960 Labour Party conference voted that Britain give up her nuclear defences – whereupon the leader of the parliamentary Labour Party said he would ignore the conference's decision. In this matter he was on the side of the Tories, and the parliamentary Labour Party has been on the side of the Tories in that and other matters ever since. At that time I was naively astonished. . . But our Labour leaders are supporting the fucking British nuclear so-called deterrent to the present day! [1]

For Gray the SSBN was only one of several reasons that supported the need for greater Scottish autonomy, but beneath his commentary on Labour's relationship with the deterrent thrives the justification for disarmament. This is a sentiment that still exists in Scotland to this day, and it should be addressed.

John Simpson of the Mountbatten Centre for International Studies has argued that the utilisation of nuclear energy can be distinguished from many other areas of modern technology by both the powerful emotional reactions and political debate that it stimulates. The beginnings of these reactions can be traced to August 1945, when atomic bombs were dropped upon Japanese cities, with dreadful consequences both for the inhabitants and their surroundings. It also resides in the responses of individuals to the intangible nature of the destructive radiation created by such explosions – something which can kill or main without being detected. Simpson also states that these concerns have been reinforced, moreover, by the walls of secrecy that have been constructed around information of their associated technologies by the governments of nuclear states for obvious security reasons. As a consequence, the public's perception of a country's nuclear aspirations is characterised by powerful national agencies, but also associated with great secrecy and the suspicion that behind the umbrella of national security there may lurk a history of immoral activities and unethical decisions.[2]

The public's anxieties over Trident have receded considerably

1 *Why Scots should rule Scotland* (Canongate Books 1997)

2 See J. Simpson *The Independent Nuclear State: The United States, Britain and the Military Atom* (London:Macmillan Press, 1986)

since 1980, and though the introduction of a replacement system in Scotland will most likely rekindle yet another heated controversy, at the time of writing public interest in the system is minimal.[3] Nevertheless, it is important to have some understanding of how Britain developed or acquired its various deterrents and why the current system resides in Scotland.

While successive British governments sought to impede the inevitable disintegration of empire, from 1945-1962 the nuclear missile submarine did not fit into UK Defence planning or its scheme for deterrence. British politics in the late 1940s and 1950s were greatly influenced by a desire for global status, focused on the need to uphold a worldwide military presence and the desire to develop an independent nuclear deterrent. The post-1945 decision to develop a nuclear deterrent was taken for two broad reasons. The Attlee government assumed the Soviet Union would have a nuclear capability by the early 1950s and believed that a British bomb would solidify the UK's position as a world power. It was also assumed that a nuclear capability represented the only effective deterrent against an opponent's nuclear, biological or chemical threat, while simultaneously defending British national interests.

To overcome the American postwar ban on information for such weaponry, after 1952 the UK began to test and deploy a number of free-falling atomic bombs with assistance from British contributors to the Manhattan Project. This was a Herculean task, and British scientists have yet to receive the credit they are due.[4] Working under stressful conditions, considering numerous designs and struggling to facilitate the adequate production of fissile materials, British administrators and weaponeers established a fully

---

3 But so is interest in politics in general.

4 During World War II Aldermaston was the site of an airfield where Spitfire fighters were assembled. It was also used by the United States Army Air Force for glider operations on D-Day. After the war, Prime Minister Clement Attlee decided to develop Britain's own nuclear weapons. Nuclear scientist Dr William Penney, who had been involved in developing the nuclear weapons at Los Alamos in New Mexico during World War II, started work and in 1950 Aldermaston became his headquarters.

independent deterrent. Testing was carried out in Australia, and several designs were considered before suitable weapons became fully operational. Yet the British still required a method of delivering its weapon.

The arrival of the jet bomber, with its ability to deliver devastating nuclear payloads, represented the first and only British model for airborne deterrence. By 1962 the Royal Air Force (RAF) successfully fielded three distinct medium-range nuclear-capable jet bombers with a range of roughly 2,500-3,000 miles. These V-bombers, named Vulcan, Victor and Valiant, involved one aircraft in each squadron being ready at fifteen minutes notice twenty-four hours a day; 30% of available aircraft ready to deploy after four hours and that figure rising to 100% after twenty hours. However, airborne deterrence involved certain disadvantages as maintenance of these strategic bombers, fuel consumption required to keep them on twenty-four hour stand-by and the safety aspects of nuclear weapons aboard aircraft were of ceaseless concern. Yet the bomber's role within Britain's nuclear defence strategy was all but extinguished in the late-1950s as the stealthy American rocket-firing nuclear-powered submarine encouraged many in the UK government to reconsider.

The postwar discovery of German wartime attempts to place newly developed missiles within submarines, and the American postwar development of thermonuclear weapons and the nuclear-powered submarine, led the US National Security Committee to approve the development of the world's first missile and subversive system, known as Regulus, in 1955. Not only had the USSR successfully deployed Sputnik in 1957, which encouraged the joint London and Washington policy against Soviet encroachment, but there was also the Kennedy administration's embarrassing cancellation of the Skybolt programme in December 1962. This system was an attempt to use a bomber to launch a ballistic missile, but the cancellation of the programme implied that the RAF was fielding a nuclear deterrent force that was increasingly outmoded. Harold Macmillan and his Conservative government were therefore provided with the necessary leverage with which to obtain the recently deployed American Polaris deterrent.

Macmillan had earlier convinced Washington to sell the UK Skybolt in exchange for the US Navy's use of the Holy Loch in Scotland for their Polaris forward operating base, a development that inspired the now legendary marches of the Scottish Campaign for Nuclear Disarmament (SCND). Due to its limitations in terms of range, a forward operating base north of the Border allowed for the Polaris SLBM to strike at the heart of the Soviet Union with greater ease. Despite considerable public disapproval the Polaris Sales Agreement was concluded by April 1963, with the British government acquiring Polaris at low cost and its missiles intended for the development of a multilateral NATO force. The people of Scotland became acquainted with both nuclear submarines and their powerful payloads by the end of the decade.

Due to the system's capabilities and consistent tensions between the US and USSR, the Royal Navy's decision to deploy its own Polaris system from Scotland was indeed controversial. A working party considered several locations for the submarine's operating base, but both geographical and operational factors dictated that construction of the Clyde Submarine Base at Faslane Bay should commence. Faslane had played an important military role since the early 1940s; the area was somewhat remote though Scotland's largest population centre, Glasgow, was within thirty miles of the facility; and the Gareloch was considered advantageous because it was sheltered and in proximity to deep water. Furthermore, Royal Naval Armament Depot (RNAD) Coulport, some thirteen miles by sea from Faslane, was constructed to perform missile servicing for the upcoming Polaris programme and Rosyth Dockyard in Fife was selected as the first British facility used for nuclear submarine refitting and refuelling operations.[5]

Work began at Faslane on 22 May 1963, the day after the government placed orders for Polaris boats, and by 1968 Polaris submarines began the first of nearly 250 patrols.

The proficient operation of the UK's Polaris fleet encouraged the Royal Navy to explore its options towards the end of the system's

---

5 RNAD Coulport will be referred to as Coulport from this point. Clyde Submarine Base, later known as H.M. Naval Base Clyde, will be referred to as Faslane.

life expectancy. The RAF formally handed over responsibility for the nuclear deterrent to the Royal Navy in late June 1969. Already in 1969, however, improvements of Soviet radar and anti-Ballistic Missiles (ABMs) resulted in a protection of Moscow which Polaris warheads could not penetrate with absolute certainty. Understandably, there were also serious concerns about the increasing effectiveness of these systems and their future in Russian strategy. Though the US initially offered the Poseidon system as a replacement for the increasingly outdated Polaris, the decision to deploy the indigenous but troubled Chevaline programme was the UK's response to Soviet technical advancements.

This system was designed to increase the likelihood of warhead survival by disguising incoming Polaris warheads as dummies to confuse Soviet ABM defences. The £4 billion programme included the need for new Polaris rocket engines; problems later surfaced with Polaris submarine hulls, and it did not provide for a Multiple Independent Re-entry Vehicle (MIRV). A MIRV capability was considered a technological leap forward because it provided the ability for a system to target each warhead independently onto separate targets. Using MIRVs, a single missile could strike several targets, or fewer targets redundantly. However, the UK's improved Polaris boats would not begin patrols with the Chevaline system until 1982. By contrast, the US Navy launched its first Poseidon boat in 1968, equipped with the highly desirable MIRV capability.

The Ministry of Defence (MoD) did not acquire Poseidon. The system was initially received with some interest, but by 1973 the American and British governments decided that it had no place in the Royal Navy as a deal could have threatened to undo the process of détente that was well underway between Moscow and Washington.[6] The improving relationship between superpowers did not last, and during the latter half of the 1970s the Soviet Union unveiled new solid-fuel SS-20 missiles. This particular

6 For the Americans détente was seen as a way of managing the Russians and the Kremlin viewed it as an acknowledgement of their superpower status. A deal with the UK could have jeopardised the state of these relations. Détente had stalled after the spring of 1976, when the two superpowers experienced complications over issues such as human rights, disarmament and arms control.

weapon appeared to represent a means of waging a limited nuclear war in Europe by taking out airfields, bases, and cities in a pre-emptive first strike. Because of the SS-20's capabilities it inspired the US to offer the UK the option of purchasing the new Trident system in 1980 as a replacement deterrent for the near obsolete Polaris. With the Chevaline improvement programme coming at considerable cost and lacking a true MIRV capability, the recently elected Conservative government, led by Margaret Thatcher, opted to renew its status as a patron of the Pentagon.

In the midst of the Cold War a replacement deterrent had to be able to pose a threat to the Soviet Union at least as credible as that initially suggested by Polaris. Ground launched cruise missiles (GLCMs) were seriously considered but these possessed several inadequacies – the lack of stealth of a sea-borne deterrent among them – and by late September 1979 Britain discarded this option because the system was considered to be too vulnerable to attack. From an objector's standpoint, central government apparently made the right decision, at least to some extent. Cruise missiles deployed by the US as part of a NATO deterrent capability also proved highly contentious throughout England as some 120,000 activists marched in opposition to the American GLCM presence at Greenham Common in April and December 1983.

A diverse combination of factors detracted from the GLCM's appeal and placed greater emphasis on Trident as the system essentially represented a Polaris force with greater capabilities and therefore improved security in terms of deterrence. The SSBN is a platform of stealth. Trident also possessed a MIRV capability that gave it a high probability of defeating an ABM system. General satisfaction obtained through Polaris since 1963 strengthened arguments for Trident, and the decision was to be finalised in the summer of 1980, with arrangements made to systematically phase out Polaris while simultaneously establishing Trident on the Clyde.

Margaret Thatcher's government was ultimately forced to choose between two different Trident systems. In 1980 the Trident I C4 was considered by the Thatcher government to be the most up-to-date and credible deterrent. By purchasing the system from the US, the UK could be spared the burden of maintaining an expensive improvement programme similar to that of Chevaline. Letters exchanged between Thatcher and President Jimmy Carter on 15 July 1980 led to the UK's purchase of Trident in the midst of a series of negotiations. Yet in 1981 the Reagan administration opted to enter full development of the Trident II D5 for the US Navy. Defence Secretary John Nott was forced to study the implications.

In 1981 the UK government presumed that if it were to accept Trident I, it would have entered service with the Royal Navy only shortly before it left service with the US Navy.[7] This implied that the UK alone would be responsible for keeping open special Trident support facilities in the US, and it would force the UK to fund any research and development needed to counter further Soviet ABM defences. Accordingly, the UK entered into agreement with the US on 11 March 1982 to purchase Trident II as described in Command Paper 8517.[8] It was this system that was ultimately established at Faslane.

The initial plan for Britain's Trident system was very different from what actually transpired however, as central government decided that its missiles should be serviced in the United States rather than at Coulport, and refit operations at Rosyth Royal Dockyard were transferred to Plymouth. This had very serious implications for job creation in Scotland. Because of the SSBN's awesome capabilities, operational characteristics and the role it served in British, NATO and Soviet military planning, the most powerful submarine weapons platform in the world was to be based in the west of Scotland with noticeable apprehension.

---

7 This observation now appears to be incorrect as the C4 system will serve with the US Navy Pacific fleet until at least 2007, when the second Strategic Arms Reduction Treaty will reduce the number of US Ohio class Trident boats from 18 to 14. Hutchinson, *Submarines, War Beneath the Waves: from 1776 to the present day* (London: Harper Collins, 2001)

8 Trident II (D5) will be referred to as simply Trident from this point. Trident II was more sophisticated than Trident I (C4) with a significantly greater payload capability. *The British Strategic Nuclear Force* Cmd. 8517 (London: HMSO, 11 March 1982)

The Vanguard class Trident submarine was an SSBN based on the British Trafalgar class nuclear-powered attack submarine design, incorporating a centrally-situated missile compartment based on a scaled-down version of the missile compartment of the American Ohio class SSBN. It was equipped with BAE Systems bow, flank, active intercept and towed array sonars, attack and search periscopes from Thales (formerly Pilkington) Optronics, collision avoidance radar and sophisticated electronic warfare equipment. While the US missile compartment had twenty-four missile tubes the British boat was to hold sixteen, as well as four torpedo tubes. Each boat was to be powered by the latest design Rolls Royce pressurised water reactor, giving the submarine a top speed of around 25 knots. Four Trident submarines would replace the previous Polaris boats, and the Thatcher government retained the option to construct an additional fifth submarine costing, at 1980 estimates, an additional £600 million including missiles. At that time the system was scheduled to remain in service until at least the year 2020.

The Trident SLBM was a three stage, solid propellant, inertially guided fleet ballistic missile with greater range and improved accuracy over Polaris, capable of carrying a maximum of twelve warheads per missile. The missiles were fuelled by Nitrate Ester Plasticized Polyethylene Glycol, a solid fuel, and had a range greater than 7,400 km (4,000 nautical miles, 4,800 statute miles) at full payload, or 11,000 km at reduced payload.

When launched, the missile was ejected from the submarine by high-pressure gas, and after it reached the surface the first rocket stage fired automatically. While in flight, the missile's inertial guidance system calculated flight behaviour and guidance. After the third rocket motor separated, the warhead carrier employed a star-sighting to confirm position and maneuvered to the point at which the warheads could be released to freefall onto the target. Using Navstar Global Positioning System satellite receivers, a circular error probable (CEP) of ninety metres could be established, making it a genuine hard target attack system with a range enabling it to hit any target in the world from any ocean.

Every boat potentially carried 192 British MIRVed warheads similar to the American W76, each with a yield of 100 to 120 kilotons, and estimated to be many times more powerful than the bomb that destroyed Hiroshima.

Now that a map of the British deterrent's journey to Scotland has been established, it is necessary to present some of the questions that should be addressed. The west of Scotland has served as the homeport for Trident since the arrival of the first boat, HMS *Vanguard*, but what was the cost of this system to the taxpayer and is security for the system adequate in the post-September 11 world? Secondly, if Rosyth Royal Dockyard served the Polaris system with distinction, why then was the shore establishment stripped of refitting and refuelling operations in 1993? Now that it has been relocated did construction of Trident facilities at Devonport Dockyard prove to be as cost effective as once expected?

In the early 1980s the persistent arguments of parliamentarians claimed that Trident had the ability to provide tens of thousands of jobs in Scotland, though the system was not acquired to strengthen the local economy but to ensure greater national security and status. Before anyone can make a proper evaluation of Trident's value to Scotland they should first have an idea about how many jobs it provided and what the total economic impact of the system has been. Yet there are other considerations.

Because the Scottish Parliament has a responsibility to address the wants and needs of the Scottish people, it is important to ask what politicians have or have not done to address the people's ongoing anxiety with Trident. Are Scotland's politicians protecting adequately the interests of the country despite limitations imposed by the Scotland Act 1998?

Knowing Scotland's experience with submarine platforms one might expect Scottish politicians to at least discuss those issues influencing employment; a discussion that is allowable regardless of their accountability to Westminster. Does the Scottish media have anything to do with this, and has it, over the years, presented clearly Scotland's experience with the nuclear deterrent?

Finally, some form of a preliminary decision on replacing the system will probably be made in 2007-2008, and certainly in the life-span of this Parliament (i.e. 2005 to 2009-10). Before we assume that a replacement deterrent will take refuge in the west of Scotland perhaps we should take interest in the possibilities. What will replace Trident?

II
The cost of security comes in many forms

ALMOST EVERYONE in Scotland should have some idea about their country's experience with nuclear weapons. But this does not appear to be the situation. Recognising the strategic or geopolitical significance of their country does not appear to rate highly among most Scots. In many ways these people have become more like that naive American cousin who is often ridiculed in the Scottish *Daily Record*, and while these visitors are often surprised by the bridge in Stirling too many Scots are astonished to hear about the presence of nuclear weapons in their lochs. At a time when NHS waiting lists are overflowing and people fear for their lives on public transport, we should be more aware of Trident's economic impact and the security of that which, we are told, makes us secure.

By 1961 Scotland was already becoming a vital fortress in the North Atlantic for both NATO and its war strategists. That year the US Navy had established a forward operating base in the Holy Loch to accommodate their Polaris ballistic missile submarine force. The Cold War had engulfed East and West, and Scottish apprehensions with submarine platforms were tempered by concerns with national security, or fears of a nuclear holocaust. Atomic blasts over Hiroshima and Nagasaki in1945, and the tension of the 1962 Cuban Missile Crisis, served as grim reminders of what could be. Responding to the international situation, the Royal Navy's own Polaris fleet eventually commissioned the 10th Submarine Squadron at the Port at Faslane, renamed Clyde Submarine Base HMS *Neptune*, in 1967. The first boat sailed in June 1968 on what was the first of numerous patrols, with support facilities at Coulport and Faslane supporting the deterrent with distinction. With American and British systems cruising Scottish waterways, people in the west of Scotland became well acquainted with both nuclear submarines and soon after their powerful payloads.

Polaris marked only the beginning of Scotland's experience with the SSBN and SLBM, and after a decade of service an ageing system that was falling behind in technological terms encouraged central government to contemplate its replacement. After Labour's colossal defeat in 1979, Margaret Thatcher and her Conservative government opted to phase out Polaris while acquiring an upgraded Trident system in March 1982. Before the party embraced unilateral disarmament, James Callaghan's government was also in negotiations with the Americans and prepared to do the same; after its downfall Labour was not given the opportunity to act. Labour in London was wholly disinterested in what Scots thought of a replacement deterrent, and, even if Scots were made aware of the situation, Labour in Scotland was generally immobilised.[9]

For Strathclyde the situation would remain much the same. Due to Coulport and Faslane's successful interaction with Polaris, both sites were to retain their status and undergo expansion in order to accommodate the significantly larger Trident programme.

With Scottish manufacturing jobs leaning towards extinction Thatcher portrayed Trident as an opportunity for families struggling in Scotland. If based on its physical appearance alone, it seemed projects for the national deterrent were indeed the answer to the country's gasping economy. One cannot overstate the massive undertaking that the Trident Works Programme represented as it was similar in complexity to both the extraction of North Sea Oil

9 The purchase of Trident was contrary to Labour's policy at that time, and within its 1979 manifesto the party took the line that Polaris should not be replaced. For obvious reasons, matters under the shady umbrella of national security are strictly withheld from the public arena. Yet with both the Scottish referendum of 1979 and a general election drawing near, it likely that this arrangement was concealed from the electorate so that Trident would not assist the SNP in the midst of a vote on devolution and have no bearing on the return of a Labour government.

and the Channel Tunnel works, requiring thirteen years to complete involving roughly 110 separate projects.

Development of the Northern Area at Faslane and work at Coulport commenced in September 1985. However, by June 1989 the estimates for works at Coulport had risen by £53 million (from £222 million to £275 million) and works at Faslane had increased by £72 million (from £267 million to £339 million), with much of the £125 million upsurge in cost attributed to pay and unexpected additions to the Programme.[10] By mid-1991 key elements in the project were roughly two years behind schedule and the scheme was reported to be running at more than double the original estimates.

The reasons for delay were numerous, but ever-improving nuclear safety standards hounded several projects causing unexpected complication. Delays surfaced with the explosives handling jetty, a facility that permitted British-manufactured warheads to be joined with Trident's missiles while aboard the boat. Another principal site experiencing complication was the Northern Utilities Building, which provided an independent power supply to sustain reactor instrumentation, control and cooling systems while boats sat in berth, and power to the facility in the event of national grid failure.[11] The project's numerous suspensions proved to be costly for the taxpayer. However, in order to be ready to support the operational programme for Trident, facilities had to be operational by a fairly rigid deadline. Combine this with, for example, the novel use of shiplift technology and fairly common civil engineering problems, such as unexpected ground conditions, and the time pressures mounted.[12] Yet the spending-spree did not stop here.

---

10 Fifth Report from the Defence Committee, 'The Progress of the Trident Programme', HC 374 of Session 1988-89

11 Second Report from the Defence Committee, 'The Progress of the Trident Programme', HC 297 of Session 1993-94.

12 Shiplifts had not been used to lift 'dirty' reactors. Vickers Shipbuilding and Engineering Ltd. in Barrow-in-Furness, which assembled the Vanguard class, employed a shiplift but only handled new reactors that did not have a decay heat removal problem to deal with. Thus, when deadlines became an issue, projects became more expensive and were more prone not to meet original specifications.

Complications with Faslane's shiplift were particularly troublesome. Built by Cementation Construction Ltd, this complex, sheltered apparatus was designed to elevate SSBNs from the Gareloch to provide the necessary system maintenance that the boat's nuclear reactors demanded over the course of their life. The powerful lift rested on approximately 600 vertical piles and 188 raker piles, which were essentially hollow steel tubes set in concrete sockets and drilled into the bedrock. Problems with the lift surfaced when seventy-one raker piles were found defective. By May 1989 half of the defective piles were removed, one-third was replaced and there were ongoing concerns with corrosion. However, by July 1992 the lift was considered virtually complete and it was handed over to the MoD in July 1993, two and a half years past the original completion date. By April 1996 the lift alone cost £314 million, £200 million over budget, and when the Royal Navy intended to raise HMS *Vangaurd* it first had to receive special permission from the United Kingdom Atomic Energy Authority.[13] Nevertheless, the shiplift's complications were not atypical amongst the elaborate projects at Trident's shore facilities and its various intricacies merely served as a single example amongst the many that the Property Sales Agency encountered.

Safety, adaptation, poor co-ordination and time restraints all contributed to increased expenditure. By July 1994 the majority of individual projects had been completed, although permission to operate independently still remained to be given due to rigorous safety standards. Safety case considerations arising from additional projects, and revised MoD requirements, also resulted in a number of design and safety alterations after the contract award. One example of these technical revisions involved facilities designed to handle the warheads and missiles on the dockside at Coulport. This consisted of ten separate facilities that included stores, missile

13 Raker piles were set at an angle and required to take the lateral forces of wind during submarine berthing. Eighth Report from the Defence Committee, 'The Progress of the Trident Programme', HC 286 of Session 1990-91; Fifth Report from the Defence Committee, 'The Progress of the Trident Programme', HC 374 of Session 1988-89; Ninth Report from the Defence Committee, 'The Progress of the Trident Programme', HC 237 of Session 1989-90

handling and support buildings, an integrated security system and the construction of access roads. At considerable cost upgraded facilities were later required to provide the necessary technical, maintenance and control services for Trident submarines. Poor co-ordination of these projects by the MoD led to compromised efficiency, which ultimately led to the Programme's added expense and further delay. With the Russian war machine drifting into sleep mode, was Britain's unfinished deterrent still worth completing after all this effort and expense?[14] Apparently, the answer was yes.

Similar to the construction of Scotland's new Parliament building, costs for this massive programme spiralled out of control. Because of tight timetables contracts were let with designs incomplete and contractors were able to submit claims for extensions of time, delays and disruption costs. Mounting costs to the British taxpayer were ultimately justified through the highest of safety standards, though the quality of guidance provided by the MoD, and inadequate design, inevitably produced governmental waste. Constant pressure to finish construction as quickly as possible led to numerous safety delays and cost overruns.

In the end the Trident Works Programme had cost a staggering £1.92 billion at 1994 prices, 72% more in real terms than originally expected, with a case-study by the Central Unit for Procurement criticising the Ministry for its 'mismanagement on a grand scale' after the cost of new facilities grew by an astounding £800 million.[15] In 1997 Scotland's annual contribution to the system was roughly £135 million and growing. At 1998 prices, acquisition of Trident cost £12.52 billion, with running costs of around £700 million per

---

14 Questions about its post-Cold War mission still hound the deterrent. The Trident system was originally intended to be the strategic nuclear force for Britain but after the MoD decided to phase-out its WE-177 tactical bombs (employed by Tornado strike aircraft) a number of warheads for Trident had to be adapted to produce a lower explosive force. They could then be used in a tactical or 'sub-strategic' role. However, this tactical role is continuously scrutinised. Questions concerning its targeting, the restrictions of its use under international law and Britain's ability to 'over-hit' are some of the more debated topics.

15 Central Unit for Procurement, 'The Trident Works Programme: A Case Study' (CUP Guidance: 1995)

year.[16] Finally, at 2005 estimates, the programme cost British taxpayers nearly £15 billion![17]

In retrospect, the Polaris programme saved roughly £45 million in capital costs by not proceeding with a fifth boat, and the overall timetable for Polaris was kept. In contrast to the development of Polaris the performance of the Trident Works Programme, though successful in its completion, was handled poorly. Membership in thc nuclcar club had its dues, and Britain was prepared to pay through the nose. However, with this membership came certain unavoidable responsibilities.

It must be understood that specifics on safety practices during service operations for Trident were not intended for public consumption and much of the current information is based on what is presently available in the public domain. With al-Qaeda's largely autonomous groups and cells performing suicide missions in London, Madrid, New York City and elsewhere, the potential for acts of terrorism involving either nuclear materials or hardware is one reason for such secrecy. Nevertheless, with the procurement, construction and operation phase of any complex nuclear weapon system came a long list of safety concerns for both the governments that employed these weapons and those communities within close proximity.

Both the Trident Works Programme and the day-to-day operation of the SSBN and SLBM naturally coincided with Scottish concerns over environmental and public health issues. In environmental terms, Trident's presence within the Strathclyde Region initially brought several inflammatory developments. In the interests of both security and operational preparedness, seemingly endless lines of security fences, penetrating observation towers, extensive geo-physical alterations to the surrounding landscape and waterways, unsightly support facilities and frequent construction traffic made for perpetual criticism. In terms of public

16 According to Geoff Hoon (Secretary of State, Ministry of Defence), if all expenditure, past and projected, is brought up to this current year's economic conditions (the non-hybrid estimate) the estimate is £14,893 million (18 January 2005)

17 Trident, Parliamentary Debates (Hansard) Vol 240 Col 778, 20 July 1998

risk, the nuclear dimension of Trident brought with it considerations regarding Soviet targeting (before the Kremlin lost its grip on Eastern Europe), ecological devastation, radiological contamination and fears of an accidental detonation involving a Trident warhead on site. All these issues were of significant concern to local authorities regardless of probability, but several issues may not have been so apparent to the general public.

In the unforeseen event of an incoming strike or unintentional blast a rapid departure from the vicinity appeared unlikely, with only two trunk roads (A814 and A817) readily available and flowing directly into the A82 near Dumbarton. With tens of thousands of vehicles fleeing the scene of an accident or strike, traffic would ultimately bottleneck near Dumbarton with the 'walking dead' preventing access for any emergency service vehicles.[18] During conventional warfare emergency services were to remain within the area to assist the wounded and fight fires. In the event of a nuclear blast, they were to be evacuated from the area if possible, then return after the explosion to assist. This sounds feasible on paper, but, in reality, is it practical? Under these circumstances, emergency service vehicles would have been trapped with those residents fleeing the area due to highly limited access. A much longer route of escape/entry along the A811 through Stirling might have been available, but this would have proved time consuming.

Radioactive fallout derived from such a blast (or blasts) is yet another consideration, and any level of participation from emergency services would clearly be dictated by its intensity.[19] However, in February 1985 the MoD decided that the construction of additional trunk roads from Coulport and Faslane was unnecessary because Polaris never experienced an accidental explosion. The Cold War never turned hot and though the Trident warhead's safety characteristics limit tremendously the probability of an unintended

---

18 Interview with Ian Leitch, former Labour councillor for Dumbarton North (19/11/02)

19 Alpha and beta radiation can cause severe burns if left on the skin or ingested into the body through contaminated food or water. The pre-eminent sceptic could argue that during such an incident limited motorway access was beneficial for the Ministry of Defence when forced to manage unsolicited media.

blast, in this day and age both accidents and acts of terrorism cannot be discounted.

Accidents involving any class of submarine are inevitable (there are numerous examples one could provide) but because there were no reported incidents with the Vanguard class, controversy remained contained. Yet while planning for radiological spills considered clean-up operations for those boats positioned in berth, there was no assumption or analysis of the consequences of SSBNs experiencing complications on the Clyde.[20] The Clyde Area Public Safety Scheme, revised in 1982 to publicise precautions taken by the MoD, was drawn up specifically with respect to accidents with submarine reactors. The most serious accident in which radioactivity is released is the meltdown of the reactor core on board the submarine. The complexity of the modern nuclear submarine increases the risk of accidents, and the causes of the various accidents depend to a large extent on both the craftsmanship of the reactor and the situation leading up to the accident. However, accidents are what they are.

---

20 It should be noted that in July 1983, when the Fund for Constitutional Government published a report by David Kaplan entitled 'The Nuclear Navy', the US Navy prepared a response in which it stated 'there has never been a reactor accident in the history of the US Naval Propulsion Program. . . The safety of the Navy's nuclear powered warships is on record. The Navy stand unequivocally behind that record.' This statement is repeated annually when the Navy testifies before the Congress. The irony of the official definitions of nuclear weapons/reactor accidents, incidents and unexpected events is that the US Navy may categorise a minor event as an accident while the public might not. The accident record of the other nuclear powers is even less known. The Royal Navy claims to have the same 'unblemished' record as the US Navy. Yet according to internal Royal Navy statistics, there were 712 'incidents' between 1962 and 1978 involving nuclear power reactors aboard British attack and ballistic missile submarines. Of the first 435 of the 712 report incidents, 205 were caused by mechanical problems, 107 by operator error, and 123 due to electrical faults. Of the 106 reactor 'scrams', 29 were caused by operator error. See D. Kaplan *The Nuclear Navy* (Washington, DC: Fund for Constitutional Government, 1983) and W.M. Arkin and Joshua Handler *Neptune Papers No. 3: Naval Accidents 1945-1988* (London: Greenpeace Institute for Policy Studies, 1989)

In 1992 the government introduced the Public Information for Radiation Emergencies Regulations to members of the public who lived near nuclear sites. Taking a quote from a booklet released to neighbourhoods surrounding the submarine base, these 'pre-planned measures are very simple but it is recognised that co-operation by members of the public, based on prior understanding, will enhance their effectiveness'.[21] This carefully worded sales pitch is reminiscent of Protect and Survive (the much satirised booklet that tells you how to make your home and family as safe as possible under nuclear attack), and is very much designed to appease the public. Why is this necessary?

Though the situation appears to be under control, governments in the West are very much aware of their public's concerns with nuclear weapons. When contemplating the production and safety of the hydrogen bombs (and their testing) in 1954, a document by Sir Norman Brook, Secretary to the British Cabinet from 1947, stated that:

> There is the difficult problem of educating public opinion in the implications of fallout. As we must secure understanding for the new direction that must be taken in our defence planning. . . this must be done without undermining public confidence. A somewhat similar problem faces the United States and our Allies in NATO. Unless the form and the timing of the solution as proposed by each Government are concerted beforehand, the morale of the Alliance may be shaken.[22]

To this day the same principle of maintaining public support applies to Britain's nuclear deterrent, especially in the west of Scotland. Yet the 1999 Clyde Offsite Safety Plan, or even a separate independent study, has failed to consider seriously the issue of

---

21 This was focused on the needs of Rahane, Mambeg, Garelochhead, Faslane and Shandon neighbourhoods. This is a Public Information Booklet prepared by the Naval Base Commander HM Naval Base Clyde for residents 'What to do in a radiation emergency' (June 1999)

22 This information was derived from a recently declassified document at Kew. Notes prepared by Sir Norman Brook for the Prime Minister, 'Fall-out' (9 December 1954). DEFE13/45

radiological contamination derived from nuclear-powered boats sailing up the Clyde.[23] Why is this so? This may be answered with yet another question. Could it be that if a significant proportion of the electorate failed to notice then their politicians may be less inclined to react?

In Scotland a handful of determined activists have successfully boarded berthed submarines at Faslane and, on several occasions, nuclear convoys transporting SLBM warheads have been blockaded. Over the course of their journey, it is assumed that two warheads are inside special containers and transported to Coulport within a specially articulated vehicle. Under civilian and MoD police escort, convoys of three to five TCHD Mk2 lorries moved warheads from Aldermaston on a monthly basis and normally passed London on the M25, took the M1/A1 to Newcastle, then travelled either west to the A74 or north around Edinburgh with all warheads escorted through the centre of Glasgow on the M8. On their journey across Scotland the warheads often travel past the site of the Battle of Bannockburn and under the foot of Stirling Castle.

Nevertheless, after September 11 2001 or 7 July 2005 it is no longer inconceivable that a car bomb or modified speedboat could damage either a nuclear convoy en route to Coulport or a nuclear vessel moored on the Gareloch. The warhead's containment and

23 Dumbarton District Council participated in the pre-planning of the Scheme and would have supervised its execution should it have ever became necessary. From 1999 Argyll and Bute Council had responsibility for co-ordinating the preparations of the Clyde Offsite Safety Plan and for managing consultations with the public. These preparations replaced the former Clyde Area Public Safety Scheme. A Local Liaison Committee was to co-ordinate, manage and liaise with local emergency services. A correspondence with Eleanor Steel of Argyll and Bute Council revealed that after 1999, 'in order to look at the plan afresh, a multi-agency group was brought together to ensure that the plan fulfilled all the requirements of civilian agencies'. However, it should be acknowledged that this plan was designed to respond to accidents involving a submarine reactor but there were no proposals for a weapon accident. See Chalmers, M. and W. Walker *Uncharted Waters: The UK, Nuclear Weapons and the Scottish Question* (East Lothian: Tuckwell Press, 2001); Correspondence with Eleanor Steel, Argyll and Bute Council, (06/07/02)

safety features make the possibilities of a nuclear disaster highly improbable, but visions of the USS *Cole* and USS *Ashland* spring to mind and the psychological impact of such an event to Scotland and the rest of the UK would be immeasurable.[24]

Currently, armed escorts do not protect boats steering through the Clyde and heading out on patrol. The Scottish Parliament might want to speak with the MoD about the impact a suicide boat-bomber could have on vulnerable nuclear-powered ballistic missile submarines. After years at university, specialists also recognise that submarines with holes tend to sink in water.

Finally, in December 1986 local authorities believed the construction of the Glenn Fruin bypass outside Coulport posed credible risks to Scotland's public safety and national security as the road permitted public access to within a relatively few metres of the facility. Two years earlier the Irish Republican Army's attack on the Grand Hotel in Brighton nearly wiped out the Thatcher government. If Irish republican terrorists, experienced in the production of car bombs and the use of mobile mortars on trucks and pickups, chose not to strike the national deterrent should we be foolish enough to expect the same from al-Qaeda? We are told repeatedly by the politicos that September 11 changed everything. If this is the case should we then be concerned with security for the national deterrent?

Despite its opposition to Trident throughout the 1980s, a policy *volte-face* also allowed Labour to adopt Britain's nuclear deterrent in the early 1990s, though Scottish Labour, being feisty and resolute, insisted on maintaining its meaningless opposition. Regardless of its duplicitous stance on Trident, the Labour Party in Scotland did not lose support in its traditional heartlands. If based on the jingoistic rhetoric flowing from shore facilities and

---

24 Seventeen people died and thirty-nine were injured in an attack on the American guided missile destroyer, USS *Cole*, in Aden harbour, Yemen, on October 12, 2000. While the ship was refuelling offshore, a small craft with two men on board – presumed to be suicide bombers – approached the ship and exploded. In August 2005, two rockets missed the USS *Ashland*, an American naval ship docked in the port. A Jordanian soldier died when one of the missiles hit the dockside.

the House of Commons, even the Americans, ever vigilant and armed to the teeth, could sleep better knowing that Britain was prepared to engage known proliferators like Iraq. If American and British deterrents functioned as efficiently as their government's intelligence services, is it safe to assume that every pound invested in the Royal Navy's system was money well spent?[25] It appears that in a weakened state of protest a majority of Strathclyde residents have grown dismissive of SSBNs.

After the arrival of HMS *Vanguard* in October 1992, the final three boats were to be delivered to Faslane by the end of the millennium. Without question the public's interest in Trident plummeted throughout the region due in no small part to Mother Russia's implosion. Be that as it may, by 1994 the US Navy was on the verge of receiving its sixteenth Ohio class Trident SSBN. Therefore, in the opinion of the MoD the May introduction of HMS *Victorious*, the second of the Royal Navy's Vanguard class, and HMS *Vigilant* in March 1996, must have appeared as anything but excessive.

Though the relaxation of international tensions inspired disarmament groups and several sections of Scottish civil society to rekindle an interest in decommissioning, by May 1999 concern with HMS *Vengeance*, and the Trident issue overall, was superseded by Labour's delivery of greater autonomy in an Edinburgh Parliament. Despite the lackadaisical efforts of the Scottish media

25 Richard Morse, of the University of Arizona and a former Los Alamos weapons designer, has argued that British and American nuclear warheads carried by submarines may fail to detonate to their full potential if fired. He claims that the casing of the W76 nuclear warhead was too thin, and that while the warhead was designed to have a blast equal to 100-thousand tons of TNT, it was likely to be less. Joining a debate over the reliability of the W-76 a nuclear arms expert at the Massachusetts Institute of Technology has concluded that at worst, 70% of those in the nuclear arsenal would explode as designed. John D. Immele, a senior Los Alamos weapons official, said that Dr. Morse's initial criticism had been deemed potentially credible and had been carefully weighed before being rebutted. Immele argues that the W-76 had an extreme test record that showed this was not a concern. However, Immele declined to discuss the details of the weapon's testing – arguably the *sine qua non* of weapons reliability.

to inspire a noticeable reaction, thus sending a unilateralist memorandum to Westminster, Scots opted to employ Scottish parliamentary elections as a mechanism that might improve their everyday living. Defence issues concerning the national deterrent were apparently considered irrelevant to most in Scotland, but to assume that one could gauge the general public's mood towards the system without complication would be entertaining the preposterous. It is, however, reasonable to suggest that Trident was tucked away into the back of the Scottish psyche and permitted to remain in Strathclyde in relative peace. Employment for surrounding communities had much to do with this, but, in the twenty-first century, how many jobs will be supported by a programme that is vulnerable to the operational requirements of military planners?[26]

26 There is the possibility that American and British deterrents encouraged 'rogue states' to become proliferators for fear of invasion, while not deterring terrorists who are able to freely mingle amongst civilian populations. Some form of a preliminary decision on replacing it will probably be made in 2007-2008, and certainly in the life-span of this Parliament (i.e. 2005 to 2009-10).

III

Memo to Fife: thanks but no thanks

WHILE POLITICIANS in Edinburgh and London make sure we share in their delight over the jobs Trident provides in Scotland, they deliberately overlook any work the system failed to deliver. Many Labour and Conservative MPs (and MSPs) would describe this assertion as being hypercritical, but given Scotland's traditional anxieties with nuclear weapons, and the Thatcher government's relentless advertising of the trade-off linking acceptance with employment, it is an observation worth mentioning.[27]

The first significant reduction of jobs (roughly 1,500 in total) involved refurbishment at Coulport in September 1982. Unlike the Polaris system, central government decided that Trident missiles would be serviced at the United States Strategic Weapons Facility in King's Bay, Georgia for reprocessing at periodic intervals. This arrangement meant that Britain did not have to construct its own reprocessing facilities.

Yet no event in Scotland's history involving the national deterrent compared to the terrible loss Rosyth Royal Dockyard endured in June 1993. While the Tory grandee Malcolm Rifkind gave the green light for planning in January 1987, for years John Stanley, Armed Forces Minister, and the Conservative government spoon-fed Fife dockyard workers the news they wanted to hear. Like its predecessor, Trident was to be maintained at Rosyth Royal Dockyard.

27 See Chapter V. In Scotland central government promoted the system more as a vehicle for economic returns than a defence issue.

The government's 1903 decision to construct an east coast naval base and dockyard in Rosyth paved the way to the creation of a new-fashioned community with roughly 30,000 residents. Though the facility was not completed at the outbreak of World War I, it went on successfully to repair numerous capital ships, cruisers and smaller craft by 1918. Yet by 1926 the civilian workload at Rosyth was considerably reduced with the yard only used to dismantle naval and merchant ships. It would not be until World War II that it was fully reopened, with men transferred from southern dockyards to bring the facility to strength. In the postwar period there were fears that Rosyth would be permanently retired but the rise of the Cold War between East and West quickly renewed its function. Just previous to Macmillan completing the deal for Polaris with the Kennedy administration in April 1963, the Admiralty Polaris Committee had already agreed that a nuclear refitting facility would be required for Polaris SSBNs. The Civil Lord of the Admiralty, Ian Orr Ewing, also emphasised that Rosyth should refit the new sea-based deterrent as well as other nuclear-powered attack submarines. Thereafter, a suitable refit yard for the submarines was to be built at the Rosyth facility, where extensions to the yard's capacity were already underway to enable HMS *Dreadnought* and later hunter/killer submarines to be refitted. It is at this point where Rosyth was reserved to the Polaris Squadron. Selected as the first British dockyard to perform such refits, it was theoretically capable of refitting any naval vessel.

After maintaining nuclear submarines for almost two decades Dockyard workers could never be satisfied fully with working at burger bars, retail outlets or any other service industry job. This is understandable. The workload associated with each SSBN refit consisted of routine shore services to provide utilities, rigging, heavy lifting and industrial support, scheduled and corrective maintenance, and modernisation projects. Scheduled maintenance included incremental overhauls and the periodic replacement of selected equipment onboard with shore-refurbished units to preclude taking the ship out of service for shipyard overhauls. Preservation of the submarine's reactor was one task included with these responsibilities. Past experience indicates that the most high-risk work was in refuelling or defuelling the reactor, and approximately fifty different technical operations were carried out during the process. Refuelling the Polaris propulsion system required cutting a hole in the pressure hull, through which the onboard reactor's fuel rods could be withdrawn and replaced. This was a highly complex procedure that required considerable work both in advance, and after the actual act of refuelling. A reactor was most dangerous when its fuel was at the end of its service life due to the nature of the physics of fission. There is considerable work to be done to bring decay heat under control, and it was necessary to chemically clean the plant before the reactor could be opened.

Refitting SSBNs involved a massive work package, with the boat essentially stripped out and the structure of the vessel cleaned. These responsibilities included a substantial modernisation programme to keep various boat systems current with new technology. Capability updates involving upgrades or additions to the likes of weapons systems, sensors or communications equipment were performed regularly. Additionally, corrective maintenance was aimed at equipment that was damaged or displayed signs of incipient problems. There also were extensive requirements for workshop and support facilities that included the capability to support traditional mechanical, electrical, and structural engineering, along with specialised facilities to support the nuclear engineering process. Rosyth's exemplary service encouraged the MoD to announce that the site was to be expanded to accommodate the significantly larger Trident SSBN in July 1984.

There are few proposals available for the Trident refitting facility at Rosyth, as opposed to the numerous applications made for Coulport and Faslane. In the available documents for the RD57 project, two dry docks, workshops, offices and support facilities on a site at the west end of the Base are mentioned. Under RD57 one dry dock was designed to measure 190 metres in length by 30 metres in width, and the other to be 150 metres in length by 28 metres in width. The main dock cover was also designed to be an estimated 50 metres wide, 40 metres high at the highest point and a maximum of 200 metres long. This structure would not only protect against weather but would support the overhead travelling cranes necessary to perform various tasks. This project was thought to require fifteen hectares for the development, five hectares of which was existing MoD land. Support services to the complex under RD57 were also intended to involve substations, stand-by power generating facilities, a compressor house, plant rooms, a materials storage area, external plant areas and demineralised water facilities, with design and construction of the facility subject to a quality assurance programme addressing safety and reliability in all matters relating to nuclear safety.

In outline, the RD46 project was intended to prepare the entrance lock to the main basin as an emergency docking facility. This was to be a location that could accept nuclear-powered boats with short notice, and was intended for support of the operational cycle of HMS *Vanguard*. This implied that it would have needed to be ready by the time the first boat was commissioned, as it was also required for support of sea trials. The scope of work for the project included construction of a docking cradle, provision of reinforced concrete pads on which to sit very large mobile cranes and backup power generators. There were a number of other upgrades to support services as the reliability of power and other support utilities was very important for docking facilities because the Dockyard had to assume that the reactor would be contaminated with highly radioactive fission products. All these facilities were to be constructed with considerable cost to the taxpayer. Yet before this ship could set sail it was already sunk.

For almost ten years taxpayers' money was poured into Rosyth. The MoD's July 1984 announcement for the Trident facility at Rosyth suggested that the cost of RD57 would initially run about £200 million; within two years it was believed the project would run closer to £220 million.[28] Phase one of the RD57 project began in October 1987 and involved the construction of a site access from the west access road, the preparation of the contractor's working area and security fencing. Phase two of the project, which involved the construction of an access causeway, necessary excavation and dewatering, was to be finished in the summer of

28 *Dunfermline Press* 12 July 1984

1991. However, what is clear is that RD57 never got past that initial concrete pour stage, though the foundations for the project were completed by October 1993 with work costing £120 million of the predicted figure. While RD46 started on site in May 1990, all projects, including RD46, RD57 and RD66, would never reach completion. To the dismay of Dockyard management, its employees and the Fife Region, in late June 1993 central government awarded Trident refitting contracts to Devonport Dockyard in the south of England. Despite the unmatched proficiency Rosyth employees acquired through Polaris, Trident would be serviced elsewhere.

For a decade employees of the Dockyard and residents of Fife were assured that Trident refit operations were to be performed at the complex, leading communities in the area to prepare and plan for their economic futures. The actual extent of these pledges is rather surprising. After the government's intentions for Trident were announced in July 1980, Rosyth received confirmation of its responsibilities in December 1984 via John Stanley. In July 1985 he once again reiterated that the Thatcher government had made an undertaking for the refitting of SSBNs to be continued at Rosyth Dockyard. By early May 1986 the government's objectives appeared incontestable as the Ministry's proposals for the refitting facility at Rosyth were circulated throughout local government. Finally, in early 1991 the Defence Committee reported that the Government wanted Trident to be maintained at Rosyth because RD57 was the only programme that could satisfy refit dates.[29] However, confidence took a nosedive when HMS *Cochrane*, the naval base adjacent to the Dockyard, appeared on the closure list in February of that year.

Fife communities were knocked for six when it was discovered that the 'Options for Change' review confirmed that the necessity of present bases was under consideration.[30] Reductions in defence

29 Eighth Report from the Defence Committee, The Progress of the Trident Programme (London: HMSO) HC 286 of Session 1990-91

30 'Options for Change' was a restructuring of the British military in 1993, aimed at cutting defence spending following the end of the Cold War. Despite its reductions after 1991, the Base would be closed by 1996. The Trident refitting complex was considered exempt at that time and MoD was permitted to proceed with its programme.

spending during the early 1990s came without warning, and defence workers across the globe were all too often unprepared for the economic downturn brought about by the subtle nature of the post-Cold War peace. By January 1992 the government did give a commitment to the continued operation of the Base, but this was interpreted as a sign of things to come.[31] Dunfermline and Fife's local authorities believed that the outcome of considerations for the Base would also decide the future of the Dockyard. These assumptions proved correct, and Devonport Management Limited (DML) sought to create an opportunity within the competitive post-Cold War market.

After the English Dockyard made an unsolicited bid to refit Trident in July 1991, and went as far as arguing for a one dockyard solution, a vicious campaign got underway. By 1992 the Trident contract became an issue of national importance in Scotland, and during parliamentary recess in September the Major government formally announced that an industrial competition for refits would take place between the two yards. John Major did not have to honour Thatcher's promises, but he was all too aware of his party's sagging popularity in Scotland. Therefore, after it was time to leave the stage this stunning announcement was made. Previous assurances made to the Dockyard by London's talking heads proved meaningless.

Rosyth Royal Dockyard was a joint venture company owned by Babcock Thorn Limited, with Babcock the majority and Thorn the minority shareholder, heavily dependent on naval refitting but commercial work steadily increasing after privatisation. That being said, Babcock promptly reacted to this precarious situation and pushed forward with a series of revised proposals. It was reported that the MoD RD57 scheme would have cost taxpayers £500 million but Babcock felt it could complete the task for £267 million by omitting the covered dockyard and employing a high level crane with a new low level fuel handling system with relocation of both nuclear and non-nuclear buildings and services. In summary, the

31 This commitment, albeit at a reduced level, involved the restoration of Type 42 frigates and preparations for numerous service personnel that were to be redeployed to Portsmouth.

new RD57 proposal was expected to reduce capital cost; maintain the required completion date; conform to stringent nuclear safety requirements; reduce refitting costs and integrate and build on the RD57 site work already performed. At one point Babcock even looked at treating the project as a completely new site so that management could change the employment conditions for employees in order to support the changes to working practices the Dockyard needed to implement. Once Babcock's project management teams approached the MoD with their intentions, and they believed there was a genuine interest, the level of involvement escalated dramatically.

A new timetable was announced in March 1993 and management went through a cycle a number of times, each time responding to further cost reductions from Devonport. It should be noted that the management team for Rosyth smelled bullshit, and had very little confidence that DML could actually upgrade its facilities for the money they were proposing. They had been studying the planning for refits years before it reached this point. Despite all this, at the request of the MoD Babcock submitted their final proposal in mid June 1993 that sliced an estimated £70 million off its previous offer. It was the growing realisation that Babcock could reduce the refit times significantly, and also reduce costs, which led the company to suggest that it had the ability to move to a single site capable of refitting the entire nuclear fleet.

The final proposal argued that Rosyth could accommodate from an early date the refitting, repair and decommissioning programme for all nuclear submarines with some dockings conducted at Faslane. Substantial work had already been completed on the preparation of a nuclear safety case for the new facilities, and Babcock was prepared to negotiate the acquisition of existing dockyard assets, participate in the funding of Low Cost RD57 and manage the implementation of the project. Besides parking nuclear submarines at Rosyth Dockyard, all refuelling work, including the removal and replacement of radioactive materials, could be successfully completed in an upgraded facility.[32] For reasons yet to

32 Babcock Thorn Management, Rosyth Royal Dockyard: No Title (Babcock Thorn Ltd; no date)

be revealed, the MoD, senior naval officials and central government ultimately disregarded this last proposal.

Though this statement fails to describe adequately the emotional upheaval of a community, the men and women of the Fife yard were distraught by the government's decision to relocate Trident refits to Plymouth. Releasing waves of bitter resentment, accusations of deceit and industrial protest, on 24 June 1993 Malcolm Rifkind, now Secretary of State for Defence, announced that the £5 billion in refit contracts would be allocated to DML because its proposal for upgrading docks and running costs undercut Rosyth's bid by some £64 million. The questionable results of the analysis for capital costs showed that the proposal to build new docks at Rosyth cost some £369 million on top of the £100 million already spent. The proposals for the upgrading of docks in Rosyth would cost £248 million, and the proposals for Devonport would cost roughly £236 million in total. Regardless of funds already spent Rifkind believed the RD57 project was more expensive than other proposals. Bids were much closer in capital cost, but there was a difference in Devonport's favour of £12 million. With regard to operating costs, Rifkind's analysis showed that Devonport was cheaper by a total of some £52 million over the period.

Taking those factors into account without including redundancy costs, the total difference between the two bids was £64 million in favour of Devonport. Questionable analysis later became a colossal blunder. In 2003, the Public Accounts Committee revealed that Trident's relocation from Rosyth to Devonport cost £300 million more than originally projected![33] In response to Rifkind's decision, Dunfermline's local authorities reacted angrily by immediately declaring that without refitting work no support could be given for decommissioning submarines at Rosyth.

---

33 In September 2003 the Public Accounts Committee announced that Trident's relocation from Rosyth to Devonport cost £300 million more than originally projected, ultimately discrediting Rifkind's 1993 analysis. Thirty seventh Report from the Committee of Public Account, Ministry of Defence: The construction of nuclear submarine facilities at Devonport, HC636 of Session 2002-03

Decommissioning Work

This is another important subject that deserves greater attention. After Rifkind's announcement, former dockyard workers left to work in the fast-food industry could take comfort in the fact that draining deep fryers was less hazardous than the disposal of radioactive waste. Submarine decommissioning plans involved a sophisticated process where a massive quantity of nuclear waste was generated. Firstly, spent nuclear fuel unloaded from reactors of nuclear-powered submarines contained large amounts of radioactivity. Just one tonne of this material contained anywhere from 950kg-980kg of uranium 238, about 5.5kg-9.6kg of plutonium and smaller quantities of alpha-emitting isotopes.[34]

When scrapping a retired boat roughly eighteen tonnes of long-lived intermediate waste was also produced. This included the cladding on the reactor pressure vessel which was made of stainless steel and, in many instances, might require machining to remove. Separation of this from the rest of the reactor pressure vessel was a difficult process even by remote equipment, and workers would almost certainly be exposed to radiation. Furthermore, it is the most dangerous material left after the spent fuel has been extracted. During submarine decommissioning operations another fifty-six tonnes of short-lived intermediate waste was also in need of disposal. This included the reactor pressure vessel, the large thin walled primary shield tank which surrounds the pressure vessel and areas of the hull and bulkhead. Finally, roughly 140 tonnes of low-level waste was to be removed and isolated. This came from the primary coolant circuit, the hull and the submarine bulkhead next to the reactor.[35] However, once the act of breaking down a submarine was complete what is to be done with the remains?

In 1989 the MoD told the House of Commons Defence Committee that by the end of the millennium a total of eight submarines, including HMS *Dreadnought*, would have been

34 This included neptunium, americium, curium and other transuranium radionuclides, and are produced in minute amounts in the radioactive decay of uranium.

35 These are only approximations and the outline of these figures was given in a parliamentary answer on 14 Feb 1997; Jaroshinskaya, A.A. *Nuclear Encyclopaedia* (Moscow:1996)

scrapped at either the Rosyth or Devonport facilities. Yet in January 1990 the Royal Navy discovered that there was a generic defect which affected the propulsion systems in all Churchill and Resolution class submarines. Polaris SSBNs remained in service, but five other boats were scrapped during the early 1990s as a result of these reactor problems.[36] They were later joined by the four Resolution class boats. By 1997 there were eleven decommissioned submarines in total, with seven boats at Rosyth and four at the English facility. By 2003 another sixteen nuclear submarines were expected to be decommissioned over the coming years. But how should any nuclear weapon state properly store or dispose of the various nuclear components and materials derived from a retired boat? The MoD initially intended to fill the reactor compartment with a solidifying matter and sink the decommissioned vessels somewhere in the north-east Atlantic. Yet this course of action was banned by the London Dumping Convention in 1983.

In 1988 the Ministry continued to view dumping at sea as a sensible option, but eight years later, in June 1996, Defence Secretary Michael Portillo announced that decommissioned submarine hulks would remain at Rosyth and Devonport until 2012. Portillo indicated that scrapped boats would then be cut into small packages which would then be stored in a deep repository. Anyone in Fife who expects these boats to removed from the Dockyard anytime soon should realise that a long-term solution has yet to be established. Submarine decommissioning plans still remain incomplete at time of writing (2005).

After their miserable interaction with Rifkind and Major, it is understandable that Fife's trust in the government's judgment was irreparably damaged. Furthermore, when the House of Commons Defence Committee scrutinised the decision, it considered the process by which management were asked to make proposals for refits as flawed and potentially unfair. Ultimately, the Committee could not confirm that the decision reached was a prudent one. The MoD had insisted on Rosyth continuing to develop RD57 whilst Devonport worked up in detail its refurbishment plans,

36 Boats included *Warspite, Churchill, Conqueror, Courageous* and *Swiftsure.*

placing the Fife yard at a serious disadvantage. The late acceptance by the government of this inequity delayed the decision by six months, but the issue was used against the Scottish yard for arguments in favour of Devonport.[37]

By 1995 rumours circulated that Trident refits might be returned to Rosyth, but if this was allowed to happen it would merely confirm that the selection was motivated by interests of the Conservative Party. After Labour's landslide victory in 1997 Defence Ministers apparently considered reversing the decision, but final details of the contractual agreement were completed by the Major government and settled just previous to the general election. This contract guaranteed that if a Labour government attempted to relocate work from England and back to Scotland it would require massive compensation to Devonport. Interment of this issue was complete. However, regardless of the potential for political bias there was other far less obvious yet highly significant factors that influenced the outcome.

Fife's local authorities were correct when they assumed that the future of both the Base and the Dockyard were intertwined, as the closure of HMS *Cochrane* was indicative of the Royal Navy's preference to be based in the south-west rather than in Scotland. In the midst of industrial competition, Babcock management also understood this to be true. The decision to refit at Devonport was also part of this situation – refits sometimes required up to two years to complete, and in that time, there was a significant naval presence at the refit location. If added into the equation that Babcock's proposal was to undertake the entire nuclear refit programme for Royal Navy submarines, not just SSBN but the entire Ship Submersible Nuclear (SSN) fleet as well, there remained a very difficult case to make in order to keep Devonport open. If the southern Dockyard was retired as a refit centre the situation became even more difficult to keep open the equivalent operational base for surface ships at Devonport, particularly given the spare

37 They also considered that a decision made on the basis of a relatively small cost difference as 'unsafe'. Seventh Report from the Defence Committee,'The Royal Dockyards' (London: HMSO), HC 829 of Session 1992-93

capacity at Portsmouth. Furthermore, Devonport had a rich naval tradition as its origins dated back to 1691, when William of Orange commissioned the building of a new dockyard to support the Royal Navy in the Western Approaches. Because of the sentimentality attached to the southern yard there were influential people in the Royal Navy who wanted this work given to Devonport.

Due to the Royal Navy's vast presence in Plymouth, it was fair to say that senior navy officials supported Rosyth but there was a significant number who pushed for Devonport. It was therefore a combination of these factors that sent Trident refits, and Fife's socio-economic situation, south. When authorities consider Trident's replacement and begin their whirlwind campaign to convince the public of their decision, politicians in Edinburgh and London will say a number of things to win support. Someone might want to be certain they did their homework (and double-checked their projections in numerous fields).

IV

Jobs, jobs, jobs

ANY PERSON or political party who dares to suggest that nuclear weapons did not offer Scottish people a long-term economic solution is placing hands on triggers. Such a suggestion frequently unleashes a coordinated attack from the best of Labour or Conservative governments on those who dare criticise the national deterrent. This assault comes in many forms and in both the US and the UK some argue that government-sponsored smear campaigns are instigated within political circles, then nurtured or overtaken by an all too receptive media.

The name of an undercover CIA officer married to Joseph C. Wilson, an officer of the United States Foreign Service and later a diplomat, was released publicly, and efforts to discredit Wilson came after he criticised President George W. Bush for exaggerating (or misinterpreting) the threat of the Iraqi weapons programme in the run-up to the current Iraq war. Glasgow's own George Galloway, former Labour MP for the Hillhead (1987-1997) and Kelvin (1997-2005) constituencies, was able to slap down the *Christian Science Monitor* after he proved that shady documents obtained in Iraq, which alleged that Saddam's regime paid him vast sums to promote its interests in the West, were forged.

Most public authorities would scoff at any such connection, but Wilson and Galloway might claim to have first-hand experience of this process. Nevertheless, of the four main parties competing in Scottish electoral politics it has been the Scottish National Party (SNP) that has been shredded in the Scottish media over its questioning of the national deterrent. It also appears that criticism of any party that supports disarmament has intensified since the end of the Cold War, and this may have something to do with Trident's vulnerability in an age where its sub-strategic role is of questionable value; especially in the war on terror. While the uninitiated reader in Scotland might presume that nuclear weapons are the centrepiece of the Scottish economy, Strathclyde and Fife's experience with Trident should dispel any false images.

Thirty years previous to the Thatcher government's decision to acquire Trident the Strathclyde economy struggled desperately despite the production of Oberon class conventional submarines at Greenock and the introduction of Resolution class boats throughout the 1970s. Therefore, it appears unreasonable for one to expect Trident to bring forth some unforeseen upsurge in the Scottish economy though, at that time, the Conservatives argued otherwise.

If we first look back to the early 1900s, the government requested John Brown in Clydebank to assemble three submarines for the Royal Navy, but after completion of these boats the yard rejected further offers for submersibles and concentrated its efforts on warships. For the next fifty years construction of surface vessels remained John Brown's bread and butter. However, with the coal industry and shipbuilding suffering rapid contraction by the late 1950s, people requiring economic assistance doubled between 1962-1972 due to the effects of high unemployment, an ageing population, and the increased percentage of single mothers and divorced wives. By this time Scotland discovered that, like the rest of Britain, its outmoded industrial plants could not vie with contemporary commercial production from overseas. Scottish shipbuilding industries were the most apparent failure, and the immense Clydeside shipyards that once delivered massive ocean liners became insolvent as several commercial enterprises once controlled by Scots had been merged into English or multinational conglomerates.

With the Scottish economy in turmoil, by July 1980 the massively expensive Trident programme permitted central government to present the economic benefits of the system to troubled Strathclyde residents in the hope that it would generate greater public support for the government, its nuclear defence policy and the system itself.

The level of employment Trident could provide for Strathclyde's workers did not rectify the Scottish economy's deep-seated structural problems, but its ability to slow the rising rate of unemployment, however limited, could not be discounted. During the operation of Polaris the civilian complement at shore establishments in June 1969 accounted for roughly 1,800 employees. However, since 1979 the Strathclyde Region had lost 154,000 jobs, of which 134,000 had been in the manufacturing sector, and in the Dumbarton district alone the unemployment rate in 1981 was 17.9% with a total of some 5,400 jobless. In 1983 the Strathclyde Region still had a 17.4% unemployment rate, and by 1984 its sagging economy provided the MoD with a golden opportunity to prove Trident's socio-economic worth as areas like the Vale of Leven, Dumbarton and Helensburgh had 6,700 unemployed in February 1984.

By this time a strapping dependency on the defence industry was made most apparent when many in Helensburgh's professional community argued that Coulport and Faslane had insulated the local area from the recent unemployment trends. Yet what was beneficial for Helensburgh was less helpful to surrounding communities; a concept that too many in Scotland have yet to discover. With unemployment in Scotland still at 14.8% in 1987, two years later its industrial workforce was only 58% of what it was in 1972. Of the top ten unemployment black spot areas in 1989, Strathclyde, Fife and the Western Isles occupied the top three places, representing more than half of Scotland's populace. Left with few alternatives, the working people of Strathclyde were forced to accept any opportunities provided by the Trident project.

Despite the persistent arguments of Conservative MPs that Trident had the ability to provide tens of thousands of jobs in Scotland, it must be remembered that the system was not acquired by central government to strengthen the Strathclyde economy but to ensure greater national security and status. Many in the west of Scotland may have misinterpreted the rhetoric produced in both local government and the Commons, and assumed that the majority of these jobs were destined for Strathclyde. Nothing could be further from the truth.

Also of significance, Trident itself was not a straightforward duplication of Polaris and, amongst a number of significant improvements, its refit schedules were less frequent and its missiles were to be serviced outside the United Kingdom. This implied that the system demanded less from the MoD but provided more in terms of performance. This leads to questions about the system's requirements and the consequent potential for a reduced civilian workforce. Because the Vanguard class incorporated tremendous technical improvements, these improvements held notable implications for shore facilities. With less service time required for the modern Trident system, various items were stripped down, cleaned or replaced less frequently, and minimal upgrading would be required during the earliest stages of its operation. Workshop and support facilities would also have less extensive demands placed upon them, and, though unclear, it appears that only accommodation facilities aboard these boats would be overhauled as frequently as that of Polaris. In terms of employment, Trident was not designed to support the same numbers as its predecessor.

By May 1984 Trident boat designs were approved, construction of the Vanguard class at VSEL was set to commence and the MoD was prepared to launch the Trident Works Programme. After releasing its Notice of Proposed Development the Ministry opted to provide employment figures for peak years of construction, rather than predicting the number of jobs the system would generate on an annual basis. This presented these figures in the most favourable light. It estimated that direct employment from Trident involved roughly 1,550 jobs at the height of construction in 1988, with 1,100 of those positions filled by Strathclyde residents and 390 to be filled by Dumbarton residents. It also stressed that a rapid build-up of construction workers was expected between 1985 and 1988, followed by an almost equally rapid decline from 1988 to 1992.

To offset the decline in numbers employed, Ministry projections suggested that numbers of Service and civilian personnel at Faslane would rise to about 800 by 1990 and to 1,300 by 1994, thereafter falling back to 500 more than the 1984 level, and remaining at this figure in the longer term.[38] Confused? This was placing heavy spin on an analysis that cloaked unimpressive new employment statistics, and the loss of 1,500 jobs at Coulport.

The Ministry of Defence had in 1984 stressed the importance of Faslane as a source of regional employment, but practically no new jobs had been created as a result of the decision to opt for Trident. British companies generally failed to win major Trident-related orders in the US, and outside contractors eventually detracted from Strathclyde as the Property Sales Agency awarded massive, specialised contracts to companies like Hawker Siddeley Power Engineering Ltd of Loughborough. The SSBN was a source of considerable expenditure on the Clyde and as a consequence, of various forms of short-term employment, but it was not liable to be a source of new long-term employment. Like that of North Sea oil, Scottish participation with Trident represented a failure of the Scottish economy to take its rightful place in the development and production of certain submarine technologies as there had been no investment in a large and powerful indigenous sector that served the SSBN.

This brings us to a comparison between Scotland's experience with North Sea oil and the country's failure to assume an active role in the production of nuclear-powered submarines. It was a sequence of events in the 1950s and early 1960s that led to the exploration and extraction of what the nationalists referred to as 'Scotland's oil'. Roughly a decade after World War II American dominance of the world energy market was under threat when Japan, Germany, Italy and France emerged as industrial rivals to the US. These were front-line states in the Cold War but because they sought to release themselves of US-controlled energy they turned to the Middle East, an area considered politically unstable by thc Americans, and the Soviet Union, which was now supplying oil technology. Furthermore, both the American government and US oil companies became somewhat apprehensive when the

38 Ministry of Defence and the Property Sales Agency 'Proposed Development at the Clyde Submarine Base Faslane and Coulport: Environmental Impact Assessment' (May 1984)

Organisation for Petroleum Exporting Countries was formed in 1960. To compensate for these developments, exploration by the US oil industry was to be focused on politically stable countries, or, arguably, with those who would not bolster the Soviet Union's own military-industrial complex through the purchase of oil technologies. Therefore, the discovery of North Sea oil was not an accident; it was both a response to demand and largely a product of strategic planning.

After the discovery of North Sea oil in 1969, the extraction occurred at the fastest rate possible. As with SSBN development and production, Scottish industry was not prepared for the oil industry, and with only a narrow window of opportunity it faced insurmountable barriers. For offshore drilling rigs, the steel required for platform construction and tubular pipelines was of a totally different character from that which the Scottish mills were capable of producing.

Comparable to platform construction, Scottish yards had no previous experience with the design or production of SSBNs while VSEL in Barrow-in-Furness and Cammell Laird in Birkenhead had already assembled Polaris boats. Furthermore, Clyde shipyards proved unadaptable and incapable of offering suitable venues for offshore platform jackets. Ships were built primarily from welded steel plate, but platform jackets were constructed from rolled and tubular steel, connected by a technique requiring specialised rolling machines and automatic welding.

With the Vanguard class submarine, the Clyde shipyards did not possess the appropriate, modernised facilities where SSBNs could be assembled on cradles and have a rail system where the vessel, in whole or parts, could be moved out to a shiplift. At several levels, the comparisons between North Sea oil and the SSBN are noteworthy. Ultimately, Scottish industry did not understand the oil industry, SSBN production or their requirements until the opportunity had already lapsed. Though Dounreay's Vulcan Naval Reactor Test Establishment in Caithness still provides a test bed for the UK's nuclear submarine reactors, it also contributed nothing to Scotland's role in submarine production.

With the somewhat unexpected demise of the communist

superpower in 1991 Trident submarines became less symbolic. The former Soviet Union had ceased to exist but this development did not destabilise the Trident project, and after January 1995 Vanguard class boats took on a secondary 'sub-strategic' role, with a number of its SLBMs carrying reduced payloads. At some point, decisions on the long-term future of the United Kingdom's nuclear capability will have to be taken. Nevertheless, while comprehensive information of the temporary employment provided by Trident remains illusive, operation of the system itself did not represent the employment bonanza initially envisaged by the Conservatives.

In 1985, Sir Hector Monro, Conservative MP for Dumfries, insisted that 50,000 jobs could have been lost in Scotland if the system was not implemented.[39] Initial reactions to such a figure were relatively favourable, but there was little detailed reference made to specific locations, direct and indirect employment, or estimates pertaining to Service personnel, civilian staff or contractors. Because the sources of this figure were so vague it could be construed as misleading. What is certain is that employment estimates were rapidly changed as the transfer of missile fitting and servicing in September 1982, and the closure of Rosyth Royal Dockyard in 1993, invalidated all previous estimates.

While a definitive study on the economic impact of Trident in Scotland has yet to occur, it appears that the system's capacity for generating long-term employment and capital investment within a number of local economies was, more often than not, of limited significance. These investments included the production of submarine periscopes in Glasgow, hulls in Renfrew, and steel plates from Ravenscraig in Motherwell for Vanguard class boats. However, while the manufacture of these SSBN components made a contribution to employment in Strathclyde it should be recognised that, within the context of the Scottish economy, these jobs represented only short-term commitments.

If we disregard estimates that included Service personnel, external contractors, local economies and such and maintain focus

39 See Trident, Parliamentary Debates (Hansard), Vol 82 Col 169, July 1985; Trident, Parliamentary Debates (Hansard), Vol 78 Col 509, 10 May 1985

on the core number of long-term civilian jobs in shore facilities provided, the picture becomes more transparent. After construction started with Trident both systems (Polaris and Trident) supported roughly 3,900 civilians in 1989, with roughly the same number in 1996. In 1993 Faslane was slated for Trident deployments and it was not until after the decommissioning of Polaris that the number of civilian jobs generated by Trident finally became clear. By February 2000 John Spellar, Minister for the Armed Forces, announced that Trident shore facilities in Strathclyde had a directly employed civilian work force of only 3,300, comprising 2,547 at Faslane and approximately 753 at Coulport.[40]

Between 1996 and 1999, there was an overall reduction in civilian manpower of 18% from a project that cost British taxpayers nearly £15 billion at the 2005 estimate. With the deterrence value of Trident under increasing scrutiny after 1991 and its record for job creation less than stunning, it is debateable whether the system represented value for money for the Strathclyde economy.

There are several ways to approach this wide-ranging subject, and considerably more research is needed on this issue before one can truly define Trident's economic importance to Strathclyde. Nevertheless, during the late 1980s Scotland's industrial workforce was only half of what it was in the early 1970s, and though the national deterrent supported almost 4,000 civilians jobs in 1989 the Strathclyde Region still had the highest unemployment rate in all of Scotland at that time. Though unsubstantiated, the Scottish Trade Union Congress also claimed that if the Trident Works Programme and the system itself were scrapped, it could have created over 74,000 direct and indirect jobs for an investment of £1.3 billion, offsetting the loss of civilian jobs at Faslane.[41]

The fact that Trident still represented a considerable number of civilian jobs and indirect employment is indisputable, but well into the post-Cold War era Trident's opposition often argued that the system accounted for zero job growth, the capabilities of

40 Civilians (Faslane and Coulport), Parliamentary Debates (Hansard), Col 738W Vol 344, 21 February 2000

41 Speech by Richard Leonard, The Economic Consequences of Trident (March 1994)

irrational rogue states did not justify a British system and Scotland's annual tax contribution of at least £135 million to its operation could be better spent elsewhere.

Trident's short-term opportunities in Scotland also failed to resurrect the Ravenscraig steelworks, which was closed in 1992 despite the production of steel plates for Trident submarines. To give some perspective on Trident, in the late 1980s the Ravencraig site generated £100-150 million of business annually in the Lanarkshire economy and roughly 16,000 jobs were linked to the company. At the same time Coulport and Faslane supported a little less than a quarter of that number. The system was not acquired to strengthen the Strathclyde economy, but with central government's attempts to suppress opposition it was repeatedly sold to Scots as such.[42]

Over the course of decades Scottish workers had been conditioned to rely on the Ministry of Defence, and though Trident safeguarded the Strathclyde economy from further despair it also did nothing, whatsoever, to invigorate it. If questions remain about Trident's impact on Strathclyde's sagging economy, the Fife Region's experience speaks for itself.

Despite the Fife yard's considerable experience with maintaining nuclear submarines, unemployment in the region was rampant during the 1980s with local economies fragile and enfeebled. Polaris clearly failed to invigorate the Fife economy, and many communities surrounding the Dockyard were areas of urban deprivation. Similar to the situation in Dumbarton, Dunfermline had some 3,100 unemployed in July 1979, 3,000 jobless in July 1980 and over 3,500 unemployed by August of that year. In January 1981 massive job losses from the coal industry placed a severe strain on the local economy, offering nothing but lean times for many of Fife's families. At the industry's peak, in the 1920s, some 27,500 people were employed by the coal companies – a quarter

---

42 Civilian employees at Strathclyde's shore facilities might also expect to encounter further reductions as submarine efficiency continues to improve. The Astute class, also to be stationed at Faslane, would not require refuelling throughout their anticipated lifetime of up to thirty years.

of the Fife workforce. By the 1970s, 65% of the coal was mined in Scotland's dozen or so remaining collieries. With companies like the American owned Trane-Ltd. prepared to lay off a majority of its employees in 1981, it was not surprising to read in the local paper that 900 people were chasing four jobs.[43] In February the district had roughly 5,900 unemployed, with the male unemployment rate for Dunfermline and Inverkeithing at 8.1% and an astonishing 20.4% at Cowdenbeath. These figures are sad at best and if Polaris represented the golden age of submarines for Fife then Trident became its Great Depression.

Because Fife communities gradually came to rely on the defence industry, by July 1982 the Rosyth Dockyard Workplace Branch of the Labour Party, a left-wing shop steward movement, offered employees courses to develop policies for an alternative economic strategy. Earlier on the Branch had conducted an Economic Impact Survey, and the results were based on 775 forms from workers in Rosyth Naval Base and RNAD *Crombie*. Amongst a number of findings, results indicated that 44% of households were solely dependent on Rosyth for their income, 35% depended upon a single pay packet from Rosyth, and 14% of households faced multiple redundancies at Rosyth. However, little came from these strategy courses with the economic situation failing to improve.

By late March 1983 Dunfermline district had nearly 7,500 unemployed and it was the Rosyth complex that single-handedly shielded the region from further economic damage. Between 1973-1984 Polaris refitting at the Dockyard provided roughly 1,800 apprenticeships in the Fife region alone, and at the end of 1984 civilian employment at the complex represented up to 25% of all employment in the Dunfermline Employment Office area with 83% of all employees from Dunfermline district and 93% from within Fife. Through a combination of refitting experience and economic necessity, many considered it a well-deserved victory when Malcolm Rifkind announced in May 1987 that 1,000 construction jobs would be created by the new submarine complex at Rosyth.

After the breakdown of the Soviet Union, the implications of

43 Trane Ltd produced air conditioning, ventilation and heat transfer equipment. *Dunfermline Press* 16 January 1981

peace threatened the certainty of Trident refits remaining at Rosyth with many in Fife becoming all too aware of the connection between Base and Dockyard. Appreciation for the naval complex was also compounded by overall economic conditions throughout the country. By 1990 Scotland's heavy industries had been reduced to only three shipyards on the Clyde, the endangered steelworks complex near Ravenscraig, and collieries at Longannet and Monktonhall. Mechanical engineering output had also slumped by 28% during the 1979-1985 period. The Silicon Glen phenomenon was subdued by the policies of the multinationals that governed it; textiles and whisky were failing; and employment plunged from 605,000 in 1979 to 412,000 in 1986.

With no alternatives, the Dockyard's civilian employees naturally hoped it would enable them to avoid the economic difficulties of the outside world. While the shore establishment had a major economic impact on Dunfermline and Fife in January 1991, at this time it only accounted for 17% of all employees in Dunfermline district. Yet in July of that year Babcock Thorn Limited, the Dockyard's management, announced 500 job losses with the government also stating that 800 civilian jobs and 1,200 Royal Navy personnel were to depart from HMS *Cochrane* by January 1992. If all civilian jobs related to the complex were lost, it was projected that unemployment in the district would rise from an already high 9.4% to well over 20%.

These horrible prospects pushed more workers to seriously consider employment alternatives to that of the SSBN, and by late 1992 it was reported that a 'promising' post-Cold War experiment in diversification of the military-industrial plant was underway to assist in developing civil business work. Whatever became of this is unclear. However, from a Scottish perspective, Rosyth trade unions believed the relocation of refitting would lead to closure of the Dockyard and discourage procurement agencies who sought contractors capable of offering one-stop shopping that was available in Scotland. Following the departure of civilian jobs from the Base the local unemployment rate shot upwards to 11.8% by November 1992, accentuating the importance of Trident refits to Fife communities. For Dockyard employees the declining economic prospects were inescapable, with viable employment alternatives largely unavailable.

Civilian employees sought to avoid defence cuts as the potential impact of Trident to Fife's local economies became increasingly evident during the final months of the refit competition. In November 1992 the impact of HMS *Cochrane* and Rosyth Dockyard to the Scottish economy was estimated to be of the order of £380 million per annum, equating to more than 14,000 jobs. Jobs related to the complex were worth around £220 million to the local economy, representing almost 10% of the regional GDP. During the 1991/92 financial year Fife council stated that an average of 4,484 persons were directly employed at the Dockyard and the total spend by the establishment on wages, materials, goods and services was roughly £169 million, £124 million of that being spent within Scotland. Should both the Base and the Dockyard be run down, estimates suggested some 18,000 people in total would be affected. Nevertheless, with 4,400 employees at Rosyth and 5,000 at Devonport, the economic importance nationally of these facilities may have been limited but to the local economies they served, they remained crucial. Like Linwood and Ravenscraig before it, the Dockyard became a symbol of Scottish industry. Waiting for the MoD to deliver a verdict would not be painless. For many, it was hoped that their children would have something still there for them.

Over the course of twenty-five years Fife's coal industry expired and the area became highly dependent on SSBN refits, yet this involvement did not shelter the region from abject poverty. By April 1993 45.5% of children at Lynburn Primary School in Dunfermline were in receipt of free school meals. Yet the far-reaching implications of relocation were also made apparent when Rosyth's commercial management, Babcock International, announced 450 redundancies at its Renfrew works near Glasgow in May 1993 and warned of thousands more job losses if it did not receive Trident refits. When the Conservative government opted to transfer refitting facilities to Plymouth, the psychological and socio-economic impact to Fife and other regions were nothing less than devastating.

Jack Dromey, national secretary for the Transport and General Workers Union, went as far as accusing the Conservative Party of placing political motivations above national interests. There were an increasing number of residents who also believed that if Trident refits were to be performed elsewhere then retired nuclear-powered boats should find sanctuary somewhere other than Rosyth. According to a Fife Council memorandum, the concept of continued basing of the strategic nuclear deterrent in Scotland was directly related to promises that the refitting of the Trident vessels would be at Rosyth.[44]

After 1993 the Dockyard did continue to refit surface ships, though at a noticeably reduced level. By late July 1995 property consultants handling the sale of HMS *Cochrane* received several bids while the Dockyard announced a £27 million contract to carry out a major refit of HMS *Newcastle*, £10 million of work on HMS *Coventry* and a £16 million order to refit HMS *Dumbarton Castle*. Though these refit orders were of significance they were simply not enough. The decision to relocate Trident operations to the south devastated Fife's communities, and though exact figures are still unclear the commonly assumed figure approached roughly 10,000 job losses.[45]

Several thousand indirect job losses were linked to local businesses in Dunfermline, there was no alternative submarine work available and valuable economic and human resources were ultimately squandered. Like other communities in Fife, Dunfermline's identity was with Rosyth Royal Dockyard. It would be some time before the future of the complex was secured by the Dockyard's proximity to the fast-expanding high tech and corporate business communities of Dunfermline's bridgehead area.

The Resolution and Vanguard class submarines can be described as Cold War investments in British security, international status and NATO strategy. To consider them a booster engine for the Scottish economy is an exaggeration. By 1999 Faslane's trade

44 Rosyth: Memorandum of Evidence from Fife Regional Council to the House of Commons Select Committee of Defence (Fife Regional Council: July 1993)

45 Interview with Rachel Squire, Labour MP for Dunfermline West (12/05/03)

unions sought protection from privatisation, and unions still experienced redundancies as rumours circulated that Faslane was under threat despite the UK's desire to maintain a nuclear deterrent. With Rosyth Royal Dockyard left to expire in 1993 and Strathclyde shore establishments directly employing a civilian work force of roughly 3,300, it is clear that both Polaris and Trident were anything but swings and roundabouts for Scotland.

The nuclear deterrent will never provide the type of inward investment that consistently delivers high employment and a positive image of the region. It will not promote international trade or the growth of knowledge industries in Scotland. There will be no corporate expansion; no guarantee that it will be in place for future generations, and it will never turn a profit. Coulport and Faslane's youngest civilian employees might wish to consider the future of SSBNs. Given the march of missile defence coupled with the decline in the nuclear deterrence role, the SLBM might not be relevant in 2025.

In 1992 the *Helensburgh Advertiser* stressed that Trident-related development had brought enormous benefits to the local area and to the west of Scotland in general.[46] The assertion was misleading. Nuclear weapons and their submarine platforms were designed to make a massive killing on the battlefield, not a constant living in Scotland.

46 *Helensburgh Advertiser* Faslane Features, summer 1992

V

Song and Dance: The Declaration of Faslane

> We, the people of Scotland,
> do hereby make it known
> that we will no longer tolerate
> nuclear weaponry on our land
> or in our waters.
> In now expressing our clear will,
> we mandate all our political representatives
> to rid Scotland of Trident.
>
> Declaration of Faslane, May 1998

LABOUR IN SCOTLAND, and other parties opposed to disarmament, appear to have adopted the sometimes outlandish and certainly unconventional methods of the self-help guru. When forced to defend their stance on Trident, these parties seem to be always releasing 'the power' of the swami. Remarkably, it appears to be working. Labour provides the best example of effective trickery. Though it has endorsed a document (above) intending to remove nuclear weapons from the west of Scotland it has made no public announcement addressing current security issues for shore establishments and continues to feed dependency on Trident by disregarding the issue altogether. Yet for one reason or another all those competing in Scottish electoral politics, including the SNP, have yet to get it right. If Trident was a hypermarket in Scotland then the credibility of these parties' policies is much like a car.

Sometimes you can lose it in the ASDA parking lot.

In the midst of the Cold War, Margaret Thatcher made known her determination to maintain British national security, and upon her arrival in Downing Street she considered it constructive when Labour's James Callaghan allowed her to observe his correspondence with Carter over Trident. Through improved security the Prime Minister intended to restore British self-confidence. However, because the Conservatives were opposed to devolution and believed there was no evidence to suggest that Scots sought change in their constitutional status, Thatcher sent shockwaves throughout the country when one of the first acts of her government was to repeal the 1978 Scotland Act. With calls for greater autonomy effectively neutralised in Scotland, there may be a correlation between Thatcher's rejection of home rule and central government's negotiations for Trident with the United States government. This is merely an assumption and will remain as such until appropriate sources throw more light on this issue. Yet while the suspension of home rule undoubtedly had much to do with traditional Conservative Party values, constitutional change in Scotland held the potential to influence an agreement over the system with the Americans.

Given the extraordinary nature of nuclear weapons it was understandable why the US government would exercise discretion with the sale of Trident. Arguably, political tranquillity in Scotland reassured the Americans of Britain's stability and this may have proved advantageous for central government in its pursuit of the system. The relationship between the two countries has not always been so harmonious. After the McMahon Act of 1946, which prohibited the United States from passing classified atomic information to any foreign country (including the UK), Britain struggled to establish its own nuclear deterrent for over a decade before securing greater interdependence with the Americans. At the end of 1949, Britain wanted full exchange of information with the US. Talks began in Washington in late 1949, but these negotiations broke down in January 1950 when physicist Klaus Fuchs, working on the British atomic bomb research project, was arrested for espionage. It was not until after the 1958 bilateral agreement was concluded, which allowed for the exchange of technical information and weapon designs, that Britain was able to re-establish the special nuclear relationship with the Americans.[47] Could Scottish aspirations for greater autonomy threaten this association? The answer to this question is unclear, but Thatcher's pursuit of the deterrent option began almost immediately.

Thatcher, being both shrewd and unwavering, understood that Scots had learned to live with Polaris and would eventually do so with its successor, especially in the midst of economic depression. Consequently, for the next five years the Conservatives sold Trident to Scotland not only on the merits of security, but through the promise of financial gain.

In mid-March 1982, Bill Walker, former Scottish Affairs Select Committee Member, suggested that Trident 'excited' Scotland with 'Scottish engineers given the opportunity to participate in the work at Coulport'. Furthermore, Sir Russell Fairgrieve, Conservative MP for Aberdeenshire West, defended the system's technological superiority over that of Polaris by arguing that the British infantry did not restrict themselves to 'single repeater rifles upon the arrival of the machine gun'.[48]

There are numerous examples of Trident being sold to Scots not as a deterrent but as a form of employment. However, the rationalisation that accompanied the Scottish public's willingness to accept opportunities afforded by the system did not necessarily translate into support for the Conservatives themselves. The Conservatives had already lost electoral ground since 1979 and Thatcher's brutal attack on state dependency came as a rude awakening in Scotland because her policies on the state's role in

47 At the end of 1941 the UK was ahead of the US in atomic bomb research, with the Americans suggesting a joint project. However, at that time Britain only wanted the exchange of information and nothing else. By 1942 the US forged ahead very quickly. The British struggled to build a few pilot units for a uranium separation plant, and only then wanted a partnership with the US. Thereafter, the British were rejected by the US, and the exchange of information ceased until the 1958 bi-lateral agreement. An agreement in 1959 also extended the exchanges of fissile materials.

48 The UK Trident Programme, Parliamentary Debates (Hansard), Vol 21 Col 29, 29 March 1982

economic regeneration came to be perceived as an assault on Scotland itself. Though the government maintained a comfortable hold on power with 42% of the British vote after the 1987 election, the Scottish Conservatives suffered tremendously as they tumbled from twenty-one seats to ten that year. Consequently, support for a Scottish Assembly was sustained as many had by this time rejected the implications of Thatcherism. Nevertheless, by the mid-1980s interest in Trident was slipping and it had become less of a defence/nuclear issue in Scotland and more of an economic question. This was a hard lesson for the political left to learn; one that took Labour nearly a decade to master and an issue that other parties intentionally disregarded.

Within a global context the threat of a nuclear exchange rapidly subsided in the early 1990s and it was this crucial factor that allowed for Scottish concerns with Trident to rest. The presence of Polaris had already outlived its controversy in Scotland, and Trident's situation became increasingly untroubled as time progressed. By this time The Campaign for a Scottish Assembly published *A Claim of Right for Scotland*, and, of limited interest to most Scots, Labour's Strathclyde European MP, Hugh McMahon, challenged the widening of Rhu Narrows for Trident.

A growing number of residents no longer actively opposed the presence of nuclear weapons in Scottish lochs, though a principled opposition to the Trident system was sustained. In October 1990 Thatcher postponed the purchase of fourteen Trident missiles for budgetary purposes, but after the Poll Tax replaced the rates in Scotland in April 1988 she was dislodged from power in November 1990. The change in premiership had no impact on the Trident programme as John Major considered it geopolitical folly to hinder Trident's progress, emphasising that the system was the minimum defence that Britain required.

The Scottish reaction to most of these developments was apathetic, and this is corroborated through yet further evidence. When the US Polaris fleet withdrew from the Holy Loch on 5 November 1991 Strathclyde residents remained largely uninspired. Relatively few recognised that the American fleet was withdrawn. Furthermore, the government's decision to order its fourth and final Trident SSBN stirred few emotions in Scotland though it left the British submarine scene vigorous. With minor complication the easing of tension between East and West, the ousting of the Scottish Conservatives and substantial progress with Trident permitted both Labour and the recently established Liberal Democrat Party to rescind their former policies on disarmament.

By May 1994 the government was forced to contemplate the days of Chevaline before an attempt to kill the US Trident programme in the House of Representatives was defeated 226-169. However, in September 1995 the SNP assumed that the international climate dictated a sensible reduction in Trident's capabilities with Alex Salmond, leader of the SNP, arguing that monies spent on an improved NHS was preferable to funds being wasted on a system with no obvious targets. Initiating a political firestorm the party wandered into hostile territory when it began to question Scotland's role in NATO and the composition of a Scottish Defence Force.

Andrew Findlay, Scottish Conservative defence spokesperson, countered the SNP's anti-Trident sentiment in April 1996 by arguing that without the system there would be no defence industry and no defence jobs. In another attempt to rouse Scottish concerns Liberal Democrat Menzies Campbell, Foreign Affairs and Defence Spokesperson, further condemned nationalist policy by arguing that 'the best you could say for the SNP proposals was that they might make sure more troops were available for the military tattoo during the Edinburgh festival'.[49] Its opposition used fear tactics to humiliate the party, with the nationalists brutally dissected over its defence policies.

Questions over Trident-sponsored employment remained. Yet their critics did not anticipate the July 1996 International Court of Justice advisory opinion ruling which stated that the threat or use of nuclear weapons violated humanitarian law (except in the more extreme circumstances). Did thirty-three years of unilateral policy finally begin to bring dividends for the SNP? A rousing moral victory for the nationalists, Greens and the disarmament movement, this historic verdict implied little in terms of votes but

49 *The Scotsman* 5 April 1996

certainly served as an embarrassment for central government and those parties recently convinced of Trident's applicability. Quandaries with local government, privatisation and other economic policies placed John Major's premiership under fire. However, in the shadow of the ICJ verdict there just might be a correlation between his visit to Faslane in August 1996 to preside over the decommissioning of HMS *Repulse*, the arrival of a fresh delivery of Trident warheads and the return of the Stone of Scone that November.

After eighteen years of Conservative rule the Labour Party was again returned to power in 1997, but by this time Trident's role in British security was assured. The unforeseen death of party leader John Smith in 1994 left Tony Blair to challenge the trade unions on Trident and complete the party's burial of disarmament, which he did successfully.[50] Menzies Campbell also argued that the national deterrent could function at a reduced level of readiness. By this time the Commons had become well stocked with Trident's advocates.

Winning a total of 418 seats, it was Labour's attacks on issues such as Conservative divisions over Europe and its promise of a referendum on Scottish devolution that had more to do with Major's defeat than the solitary issue of Trident. In Scotland alone, Labour was also able to manage a commanding fifty-six seats. Phil Gallie, Conservative MSP for the south of Scotland, believed that Trident was not a factor in the loss of support for the Tories, though Rebecca Johnson of the Institute for Disarmament Diplomacy asserted that 'the Conservatives inability to provide a stance [on Trident]', left the party flustered by its inability to 'monitor the pulse of the voting public'.[51] Most informed people would likely favour Gallie's assumption.

---

50 Disarmament had left John Smith with unpleasant memories and Conservative MPs had challenged him on jobs and the nuclear deterrent. Though the party in Scotland maintained a symbolic non-nuclear defence policy to appease its followers, it was Tony Blair who fianlly convinced the unions that opposition to Trident served only to damage Labour's credibility and so granted Major another term.

51 Interview with Phil Gallie, Conservative MSP for the south of Scotland (16/04/02) Johnson, R. British Perspectives on the future of Nuclear Weapons (Washington DC: Henry L. Stimson Centre, 1998)

By this time some Liberal Democrats believed there was no reason to maintain Trident as a status symbol. However, this should not imply a willingness to decommission as their 1997 manifesto pledged to maintain Britain's basic nuclear capability and restrict the number of warheads to the same number as previously deployed on Polaris. Like Labour, a minority of Liberal Democrat MPs, and, eventually MSPs, openly opposed Trident but by this time current policy had assisted the party in doubling its numbers.

Yet complete disarmament still had its champions. The SNP rejected Labour's devolution scheme in 1997, as a Scottish Assembly would be powerless to remove either Trident from the Clyde or decommissioned submarines from Rosyth. The Scottish Green Party was also fundamentally opposed to the billions of pounds spent on Trident, but the SNP's manifesto promised to negotiate a 'phased but complete' withdrawal of Trident and planned to invest these savings in conventional defence, health and education.[52] The nationalists gained two seats and only slightly elevated their vote in 1997, but they were now second to Labour in Scotland. With a Scottish Parliament on the foreseeable horizon they intended to pursue the Trident issue, and Labour's track record on disarmament, to the fullest extent. The results were questionable.

Labour's 1997 manifesto promised to readdress the Scottish constitutional question although some Scots MPs believed that a Labour victory should be mandate enough to deliver the Parliament, as the SNP did not represent a tactical vote with a majority of Scots voting Labour in 1979, 1983, 1987 and 1992. The referendum was therefore one of the first acts of the Blair government, and on 11 September 1997 the result demonstrated that devolution was indeed the settled will of the Scottish people.

52 The Greens would decommission Trident immediately if given the opportunity. Also see Nuclear 'Triple Whammy' facing Scotland (SNP Archives), 27 December 1996; Green Party Executive members join SNP – Party offers best way forward for Scotland (SNP Archives), 17 December 1996; The SNP believed that phased withdrawal allowed for the transition from the nuclear defence industry to an alternative source of employment. SNP, Yes we can win the best for Scotland: the SNP General Election Manifesto 1997; Scottish Green Party Election Manifesto, Scotland's Future: Green of Grey? (Edinburgh: 13 February 1997).

A total of 74.3% voted for the Parliament and 63.5% endorsed its tax-varying powers, indicating that both the middle and professional classes who had been so doubtful in 1979 were now convinced otherwise. Years of antipathy to Thatcherism and Conservative rule strengthened this sentiment, with the proportion of support effectively silencing any opposition.

Either a political masterstroke on Labour's behalf, or pure coincidence, the system was not permitted to be a factor in 1979 and the party took confidence from Scottish disinterest with Trident as the bill was published in mid-December 1997. Since there had been no major incident with nuclear hardware on the Clyde, after the Dunoon protests in the early 1960s many Scots believed acts of resistance were nothing more than acts of futility, linked acceptance with employment and came to tolerate SSBNs. However, though some have claimed that Trident was an embarrassment for the party due to its previous record on disarmament, after Scotland confirmed its desire for greater autonomy Defence Secretary George Robertson made sure to emphasise that there would be no return to the unilateralism that troubled Labour for so many years. Comfortable with its decisions and certainly aware that a solid majority of Scots still rejected independence, elections for the resurrected Parliament were scheduled for 6 May 1999.

Given the extraordinary nature of SSBN and SLBM systems it was understandable that central government would exert absolute control over their deployment and usage, even if that strained relations with Scotland. Yet in the public's eye, decommissioning often meant tearing the 'Great' out of Britain, just as the public was sometimes convinced that possession of Trident could somehow replace empire. Therefore, the intensity of the nuclear weapons question often varied according to the international climate and the state of the British economy, with Trident generally considered a necessary component of national security, foreign policy, spending plans and employment.

During the SNP's campaign for the first Scottish Parliament election the party allowed for at least one highly critical oversight with Trident. The Scotland Act 1998 (Schedule 5, Part 1, Paragraph 9 (1)) clearly indicates that Westminster retained all powers and responsibilities for all defence matters.[53] Though it was not an issue open for discussion the SNP launched a campaign against Labour, its slippery relationship with Trident and the system's immunity from Scottish objections. The SNP and the SCND had already voiced their opposition to Scotland's annual tax contribution that was spent on the system, while Alex Neil, SNP's vice convenor on policy, suggested that the Scottish Parliament should withhold funding.

The nationalists also addressed lingering employment concerns, but its strategies were confused and attention to detail was left wanting. According to Frank Purdie, Faslane Shop Steward, Lloyd Quinan, SNP candidate for Dumbarton in 1999, spoke of bringing work for trains and fishery patrol boats to Faslane as an alternative to the system, while the SNP candidate for Rosyth also promised the same. Not only was this contradictory, but, in geographical terms, it was impractical for boats from the north east of Scotland to travel to Faslane when Rosyth was closer. So seriously flawed was this policy that proposals for a Defence Diversification Laboratory and Management Division could not be taken seriously by the electorate. These types of miscalculations are inexcusable and it is clear the SNP did not do its homework.[54]

Labour in Scotland also voted to scrap Trident at its 1998 conference in Perth but it was external pressure from parties like the SNP that forced the party to re-declare its duplicitous stance to the electorate. Labour in Scotland had frequently voted for disarmament in the past, but one must understand that this range of opinion was not unique as it also persisted in England. Arguably, the divisions on this issue were sharper in Scotland for three fundamental reasons: the party in Scotland was institutionally more left; Trident was sited in Scotland; and there were alternatives in Scottish politics as the SNP and Scottish Green Party vigorously opposed the system. These factors intensified the issue for Labour.

53 Reserved matters included the defence of the realm; the naval, military, or air force of the Crown, including reserve forces; visiting forces; international headquarters and defence organisations; trading with the enemy and enemy property. Scotland Act 1998, Schedule 5, Part 1, Paragraph 9 (1)

54 Interview with Frank Purdie, Faslane Shop Steward (02/06/03)

The party therefore tended to default slightly more towards an anti-nuclear position because it was driven by other outside imperatives.

By 1998 the issue became formulaic for the Scottish party and Trident no longer resonated with voters; there were in the past routine demands for its withdrawal by Labour in Scotland but these calls were routinely ignored and Scots MPs did not expect anything more. To some extent it appears that Labour in Scotland was compelled to accommodate the Scots on this issue for its own political well-being.

The situation was similar for Liberal Democrats north of the Border, which also had leftish and Green sections, and while their opponents often referred to this as hypocrisy these parties preferred to label this as pragmatic politics. This may be where the psychological techniques of the motivational guru take hold; bolstered by a five-point plan.

First, Labour in Scotland considers the decommissioning of a system that sustains jobs in communities surrounding shore establishments. Then, London retains Trident but permits Labour in Scotland to vigorously amplify its opposition to the system and its non-nuclear defence strategy. After slinging a bit of political mud at the SNP for its position against deterrence, the Scottish party defends Trident and it never activates a jobs package that could create jobs throughout the west of Scotland. Finally, Labour in Scotland can rest; its job is done. Before you know it, and without realising it, the Scottish public is left completely spun out of its mind. It is all part of that crazy game called quasi-autonomy. Uncomfortable with its appearance, Labour's participation in the creation of the Declaration of Faslane could only make for a hollow and highly symbolic gesture.

With roughly 10,000 signatures collected, on 11 May 1998 Alex Salmond, Cathy Jamieson of Scottish Labour Party executive and the Church of Scotland launched the Declaration. However, external influences and political aspirations may have been responsible for Labour's participation and the Scotland Act 1998 ultimately left their assertion meaningless. The Scottish Parliament has no locus in a straightforward defence debate. Furthermore, it has been conveniently preoccupied with a number of domestic issues and has been in no rush to place Trident on its agenda.

In fairness Labour in Scotland had to consider civilian and Service employees who, along with families and friends, could have assembled a similar number of signatures for Trident's retention. This crucial factor, coinciding with overwhelming civic disinterest, spared Labour from further humiliation though the SNP was ahead by 10% in the polls by July 1998. Nevertheless, just months previous to the first democratic version of Scottish parliamentary elections Labour endured a battery of criticism over Trident from local and national media.

Based on the premise that there had been a relaxation of tension since the end of the Cold War, in July 1998 the government's Strategic Defence Review confirmed Labour's commitment to Trident with the system subjected to 'minor rather than fundamental' changes.[55] This may have played some part in the SNP's recent success in the polls but the release of this document previous to the Scottish elections also demonstrated the Blair government's confidence both in its decisions, and with the Scottish electorate. This assertion is bolstered by one highly significant event. Just days previous to the vote for Scotland's first democratically elected Parliament, central government was so comfortable with Trident's presence in Scotland that it actually permitted the fourth Vanguard class boat, HMS *Vengeance*, to make its maiden voyage to Loch Long. The Scottish media became interested in the issue once again, forcing Labour in Scotland to perform.

When considering Trident there are numerous examples of the print media's assault on Labour. Yet even the most acrimonious of criticisms failed to rekindle serious public interest with Trident, and results from the 6 May 1999 election confirmed this. During the parliamentary election those who supported Trident gained a majority of seats. The Liberal Democrats captured seventeen seats, the Conservatives managed eighteen, but Labour, despite its complex relationship with the system, was able to secure

55 Trident's capacity would be reduced to forty-eight warheads per boat. T. Dodd and M. Oakes 'The Strategic Defence Review White Paper' (International Affairs and Reference Section: Cmd Paper 3999)

a reasonable total of fifty-six seats within the Scottish Parliament. Those who opposed Trident took the remainder. Both the Green and newly established Scottish Socialist Party acquired one seat each, and the SNP, experiencing a swing since the 1997 election, accumulated a respectable thirty-five seats. This was a bitter disappointment for the nationalists, and their position on Trident did little, if anything, to assist them.[56]

Similar to 1999, the Trident issue would also have no relevance whatsoever in the 2003 Scottish elections. However, the Helensburgh community and civilian employees might start the push to pursue various protective measures now, rather than waiting until the last minute.

If Scotland's Labour Party is genuinely interested in protecting civilian jobs there are certain protective measures that might need to be considered (Appendix A). Without violating the guidelines set by the 1998 Scotland Act, the Scottish Parliament may have the power to evaluate periodically, and independently, Coulport and Faslane's military significance to current and projected theatre strategy.[57] 'To work together, where appropriate, on matters of mutual interest', it might also nurture an adequate awareness of developing defence technologies.[58] Operating within the limits of the 1998 Act, a formal non-partisan research committee could

---

56 Christopher Harvie has stated, 'the decrease in oil prices, farm and fish products and microchips that assisted Labour as the SNP was forced to admit that the economics of independence appeared unreliable' C. Harvie *No Gods and Precious Few Heroes: Twentieth Century Scotland* (Edinburgh: EUP, 1998)

57 Theatre strategy is the art and science of developing integrated strategic concepts and courses of action directed toward securing the objectives of national and alliance or coalition security policy and strategy by the use of force, threatened use of force, or operations not involving the use of force within a particular theatre. Just in terms of strategic planning a shift may be on the horizon. According to the current American president, the greatest danger the US faces remains with North Korea, Iran and acts of nuclear terrorism, See George W. Bush, The White House, September 17, 2002

58 See Chapter 7. Also see 'Memorandum of Understanding and supplementary agreements', which can be found at: http://www.scotland.gov.uk/library2/memorandum/default.htm.

prove beneficial because this information might be used to act as a watchdog for the civilian employees of shore establishments.

Labour in Scotland may point their fingers south and argue that this was a job for London, but by supporting a second opinion it could assist the party in securing Scotland's economic well-being. Scotland as a whole is suggested, rather than the limited area surrounding Coulport and Faslane, because evidence suggests that the public has been led to believe that Trident provided a boost to the Scottish economy. With the Scottish Parliament acting as a security net there is the possibility that such measures could protect shore establishments from the ravages of poor decision-making. It could also be argued that such precautions could prepare civilian employees for the worst case scenario, thus allowing authorities time to react appropriately. Rosyth's experience has shown that Scotland needs to be meticulous in its attention to the national deterrent though Labour in Scotland, up until this point, apparently chose to do nothing of the sort.

VI

Spin and Confusion

FOR THOSE OPPOSED to nuclear weapons there is little doubt that Trident highlighted the downside of Scotland's quasi-autonomous status, while its champions remind us of the security such systems provided throughout the Cold War. Nevertheless, there has been an incalculable number of people in Scotland who have chosen not to crash Faslane's gates yet have maintained a principled stance against nuclear weapons. With the situation as such, Scottish civil society was left diffused and incapable of establishing a dominant rallying force that worked against Trident.

It is immediately recognised that the system highlighted obligations that were linked with the role of the welfare state. This, of course, was relevant to a number of subjects, including matters of economy and foreign affairs. However, the benefits of Union definitely came with a price and accommodation of Britain's nuclear deterrent represented one such fee. Because Scots had learned to live with Polaris, both Conservative and Labour governments were granted the luxury of only occasionally having to address the Trident issue. One must also recognise that a significant number of people appreciated the opportunities the national deterrent provided, and an even greater number took pride in the contributions Scotland made to the British military establishment.

However, what level of influence can be attributed to our daily sources of information? Why after the arrival of the US Polaris force to the Holy Loch, and the Dunoon protests of 1961, did

Scots learn to restrain their concerns with nuclear weapons? [59] There are a number of reasons, but one should not discount the level of distortion that comes from sporadic media coverage or the inconclusive nature of opinion polls.

The media are crucial to the formation and preservation of national identity and when considering the Scottish media's depiction of Trident, this assertion becomes unquestionable. Most of Scotland's major newspapers have long histories, with the *Glasgow Herald* founded in 1783, the *Courier and Advertiser* in 1801 and the *Scotsman* established in 1817. The *Scotsman* lurched to the right in the 1990s when it was taken over by the Barclay Brothers, and the *Glasgow Herald* became more left-of-centre during the 1980s. Yet when attempting to gauge the influence that the media have held over Scots' views on the SSBN, one has to appreciate the not so subtle differences between local and national outlets. Let the spin begin.

Local papers are entirely driven by local perceptions. When considering Trident, their motivation was inspired by whether they believed their readers were generally motivated by either employment or concerns with the system. If there was uncertainty, in many instances these publications attempted to reflect both. Providing an example, a local paper in Peterhead would defend the fishing industry to the hilt while a local paper in Caithness would tend to support employment at the Dounreay test establishment because that was the predominant issue. Yet that same publication would also acknowledge the environmental question of Dounreay. When considering the local media's position on Trident, the *Helensburgh Advertiser*, the *Lennox Herald* and the *Dunfermline Press* made clear this assertion.

59 On 4 March 1961 SCND, working in conjunction with the Direct Action Committee Against Nuclear Weapons (DAC), assembled roughly 1,000 fellow activists who marched from the lochside community of Dunoon to the Holy Loch in opposition to its American defender. Several demonstrations were also held on 14 May, and on 20-21 May, with the latter representing an escalation of the movement's efforts. However, in the midst of the Cold War the deterrence value of Polaris, regardless of its foreign status, soothed much of Britain's post-1945 security concerns and the system remained firmly in place.

Arguably, strategic headlines relay the particular message a paper chose to deliver, regardless of agenda. In the period between 1979-1999 local newspapers for those communities heavily reliant on the employment Trident was thought to provide clearly demonstrated a push to the right. In the west of Scotland, the *Helensburgh Advertiser* proposed sustained employment, necessity and safety through headlines which read 'Future of the Bases is Secure' in July 1980, 'Our defence is at stake' in September 1981, and 'Technical safety award for Faslane' in May 1996.[60]

The situation was much the same in the east of Scotland. Before Trident's relocation the *Dunfermline Press* headlines frequently declared 'Base fit for the task' in March 1984, 'Trident task for Rosyth' in July 1984 and 'Yard expansion talks: Council seek fair deal on Trident plan' in January 1986.[61]

Local papers serving communities less accessible to Trident, especially in terms of economic advantage, tended to lean to the left and often brandished criticism. The *Lennox Herald*, also in the west of Scotland, was a prime example of this. Headlines such as 'Trident missiles a necessary evil?' in August 1980, 'Fury over Trident II Purchase' in March 1982 and 'Ministry faces flak' in February 1985 all presumed Dumbarton's dissatisfaction with the decision to replace Polaris with Trident.[62] However, there is the danger of inaccuracy if it is not recognised that local papers incorporated the arguments of both sides within individual stories and published articles, though with noticeably less frequency. In February 1981 the *Dunfermline Press* offered 'CND unfurl their banner' while the *Helensburgh Advertiser* explained that 'Defence Secretary urged to go public on problems at Faslane' in May 1989.[63] It was sporadic, obligatory entries such as these that only rarely acknowledged the Trident controversy in Dunfermline or Helensburgh.

60 *Helensburgh Advertiser* 18 July 1980; *Helensburgh Advertiser* 18 September 1981; *Helensburgh Advertiser* 30 May 1996

61 *Dunfermline Press* 25 March 1983; *Dunfermline Press* 12 July 1984; *Dunfermline Press* 7 January 1986

62 *Lennox Herald* 15 August 1980; *Lennox Herald* 12 March 1982; *Helensburgh Advertiser*, 15 February 1985

63 *Dunfermline Press* 27 February 1981; *Helensburgh Advertiser*, 20 May 1989

National broadcast and print media favours either committed or partisan journalism, with patrons of this style arguing that it makes clear where the author stands; or it tends toward neutral or detached journalism, with proponents arguing that it provides a more accurate and unbiased service to the public. This form of media is more generalised as it tends to focus more on issues of national, rather than local, importance. Nevertheless, it appears that television often disregards protest and focuses more on personalities. Vocal activists associated with the disarmament movement are often depicted as radicals, or leadership for the swarms of the 'great unwashed' that occasionally appear. If this is the case, do broadcasting networks have agendas that go beyond the search for an audience? A verdict on this has yet to be established, but perhaps the issue of ownership can shed some light.

In 1940 there were nineteen titles in Scotland, eight of which were English-owned. By 1980, the year of Trident's procurement, there were only fifteen papers with four in Scottish hands, two were English-owned and no fewer than nine were the property of multinationals. Given that modern communications tend invariably to globalisation, journalist Ian Bell argued that:

> The problem of the Scottish media is easily put down; how do you represent, report or foster a national identity – even if you are sure what it is – when everywhere communications are falling into the hands of multinationals who, by definition, have no country? Assimilation has long been a problem for the Scottish media confronted with the demands of a larger neighbour. . . To that extent, the Scottishness of the Scottish press is merely superficial, if by Scottishness it is meant a willingness to reflect the views of the majority. [64]

For some time the *Scotsman* was owned by a subsidiary of the International Thomson Organisation, Thomson Regional Newspapers. Such papers, again according to Bell, 'have for decades, perhaps for centuries, acted as a distorting mirror'. Therefore, if we focus our attention on the *Scotsman* and the *Herald* newspapers

64 I. Bell 'Publishing, Journalism and Broadcasting' in Paul H. Scott (ed.) *Scotland: A Concise Cultural History* (Edinburgh: Mainstream Publishing, 1993)

and disregard the *Daily Record*, which continued to serve the Labour Party, devolution and the largest block of voters 'with unwavering journalistic selectivity', it becomes an effort to deny his relentless criticisms.[65]

A swift overview of these two papers bears the unmistakable mark of ambiguity, which ultimately solidifies Bell's theory. In March 1982 a *Herald* headline declared 'Trident system is off target', insinuating futility, yet by May 1986 the *Herald* apparently rejoiced when it announced a '£220m project for Trident fleet refits'.[66] That same paper expressed frustration when it stated 'Clyde jobs blow as work is given to the US' in September 1982, but adopted a more defiant stance in October 1992 when it claimed 'Anti-Trident protesters brushed aside at Faslane'.[67]

The same degree of confusion applies to the newspaper that regarded itself as Scotland's national paper. In May 1995 the *Scotsman*'s headline announced 'Hopes raised for Trident work at Rosyth', suggesting support for the Polaris replacement, while it chastised the 'Deadly war machine [that] sweeps aside Faslane peace protesters' in March 1996.[68] A consistent perspective on Trident is never established at any point, and while these papers simultaneously commend and condemn the system throughout the 1980s, this journalistic imbalance is again employed throughout the 1990s. Bell acknowledged that neither paper had been found guilty of disloyalty at any point in their histories, yet 'both have lived the contradiction: Scotland's national newspapers refused to seriously question the political settlement fundamental to the United Kingdom'.[69] This same indictment should be applied to their handling of the Trident issue.

Various media techniques had also been applied to the subject of Trident in Scotland. Symbolism is contained within traditional myths, with weapon systems purposely given classical names to set

65 *Ibid*.

66 Glasgow Herald 12 March 1982, p. 1; Glasgow Herald 2 May 1986

67 *Glasgow Herald* 10 September 1982; *Glasgow Herald* 26 October 1992

68 *The Scotsman* 18 May 1995; *The Scotsman* 20 March 1996; *The Scotsman* 27 September 1997

69 Bell 'Publishing, Journalism and Broadcasting'

them in the structure of traditional culture. 'Trident' is not only the god Poseidon's weapon, but also Britannia's, and this symbolism may not be without impact on some British minds. Comparisons with accepted primitive weapons was a way in which nuclear weapons were classified as part of human culture, and thus, these weapons may be subconsciously regarded as less dangerous than they actually are.

This was a strategy employed by the defence industry, but a strategic vocabulary, or symbolism, had also been consistently applied to media presentations. According to linguist Paul Chilton, one British newspaper believed the neutron bomb 'will give Europe a shield'. Simply stated, Chilton's argument was, 'who would object to a purely defensive shield?'[70] In terms of word association, the Scottish media has more often than not associated the system with terms of defence (protection), employment (financial security) and safety (invulnerability), while activists had been typically linked to radicalism (social irresponsibility), unemployment (personal irresponsibility) and defectiveness (the term, 'the great unwashed', springs to mind yet again). Depending on one's opinion of Trident, either side of this argument could be rationally construed as a misconception.

Through certain media outlets, the responses of those concerned with Trident were made unequivocal. However, when attempting to establish a perspective on this subject a conundrum is fostered through the inaction of the silent majority. While views on the Anglo-American invasion of Iraq were amplified through nationwide protests in 2003, the same could not be said for Trident. These views were also susceptible to modification as concerns with the Cold War and the post-Cold War Trident were two very different issues altogether. Therefore, it is insufficient to theorise that a majority of Scots generally rejected these weapons simply because the disarmament movement exhibited concerns when Polaris submarines suffered cracks in their secondary cooling systems or when ballistic missiles were dropped during transport at the Holy Loch.

70 P. Chilton 'Nukespeak: nuclear language, culture and propaganda' in Crispin Aubrey (ed) *Nukespeak: the media and the bomb* (London: Comedia Publishing Group, 1982), p. 105

It is equally inaccurate to suggest that a majority of Scots generally supported procurement of these systems just because livelihoods were derived from Trident's shore facilities, especially when a significant number of employees preferred another line of work. There is little doubt that a majority of citizens held an opinion on Polaris or Trident, but an official forum was not made available and most chose not to advertise their beliefs for one reason or another. For these reasons alone it is reasonable to suggest that Trident was received in Scotland with a fluctuating degree of ambivalence, though a verifiable determination on this particular subject might only be completed sometime after the system is decommissioned or replaced.

Hesitancy with Trident nurtured divisions within Scottish civil society but it did not inspire so powerful a split that it encouraged most Scots to re-evaluate their country's status as a quasi-autonomous state. This was the most ideal situation the politicos could expect. When considering Trident's repercussions on either cultural or political identity towards the end of the Cold War (1979-1989), the system was naturally understood as a deterrent to foreign aggression. Notwithstanding this, the efforts of Scottish separatists to remove Trident from Scotland were ultimately hampered by simple lack of interest, security concerns and a struggling economy that needed jobs enough to accept the system's necessity. The fact that procurement placed further financial strains upon the national economy was not highlighted. Furthermore, it is certainly possible that, at that time, a majority in Scotland recognised that reorganisation of the UK's military framework only served to compromise national security.

So why did the post-Cold War era not prove to be more advantageous for separatists when considering Trident's impact on these identities? The answer lies with three, possibly four, justifications. From 1989-1999 Trident still employed civilians in Strathclyde, and many believed there was little point in discarding a source of employment after the threat of nuclear exchange had begun to subside. Interest in these weapons, in both societal and political terms, also plunged. Surprisingly, terrorism has yet to rekindle an overwhelming interest in nuclear weapons in Scotland. Amongst a number of problematic issues, the SNP had also not devised proposals for a military capability that was at least as credible as that which was already available.[71] In this instance Trident presented few complications for those seeking only a limited measure of home rule, as defence was a matter reserved to Westminster. However, the SNP's arguments for self-government unravelled under the weight of economic and military considerations alone, and for a majority of Scots, independence did not constitute a sound personal or political purpose.

Though Trident no longer resonated with the Scottish electorate in the post-Cold War era, it remained a lingering though diminutive issue. The military is still a source of political unity with the UK, but this does not discourage some members from voting for the SNP. It could therefore be argued that some in Scottish Regiments supported the relocation of Trident. Yet an interview with Major (retired) Alastair Campbell of the Argyll and Sutherland Highlanders may best describe the outlook of those serving who did not share this sentiment. Campbell stated that:

> . . . the considerable majority of those in the forces have no sympathy with the peace movement, or any of its tenets such as unilateral disarmament. In the forces, generally, we like to think of a strong deterrent to ensure that we do not have to fight, especially the nuclear battle which is too horrible to contemplate. There are some who believe in the policy of the SNP – but not very many, certainly not a 'significant number'.[72]

At time of writing (2005) British troops are serving in Iraq regardless of Britain's possession of Trident, but in all likelihood, Campbell's view was the opinion of the majority. Yet many Scots, including those with no affiliation with the nationalists, have taken a principled stance against nuclear weapons. This was made evident by Labour's meaningless policy in Scotland, which consistently

---

71 See J. Hawthorn: Some Thoughts on an Independent Scottish Defence Force: Occasional Paper No. 1 (The Scottish Centre for War Studies, University of Glasgow, 1998)

72 Interview with Major (rtd) Alastair Campbell of the Argyll and Sutherland Highland Regiment (12/08/03)

maintained a non-nuclear defence strategy despite the determinations of the London leadership.

In 1999 the Scottish Parliament's public petitions committee was obligated to consider Trident, though the Scotland Act reinforced central government's monopoly on defence. In December of that year, one petition called for Parliament to conduct a review of its obligations under both Scots and international law in relation to Trident's presence. Another requested that it reviewed the relevance of the International Court of Justice verdict on the legality of Trident in Scotland, and a third hoped to establish that Faslane contravened the ICJ finding regarding the threat or use of nuclear weapons.[73] Because Edinburgh wants to avoid any trouble with London these petitions were ultimately rejected. Yet they served as a reminder that Trident was still a matter for some concern in Scotland.

If these examples still failed to bolster our understanding of the Scottish perspective on Trident, public surveys were also of limited assistance.

Opinion polls were not subject to verifiable proof and were commonly represented as value judgements, preferences, or estimates of the outcome of future events. Most politicians regarded attitudes as more fundamental generalised predispositions, opinions as specific manifestations of underlying attitudes and values as people's ideals and the commitments they make to pursue them. Opinion polls are not precision instruments, although they are frequently taken by newspapers and their readers to be such. Polls are, however, fairly basic devices for charting the general development in public opinion as it descends upon the major issues of the day. Though the opinions of Scots on nuclear weapons have been charted over a twenty-year period, it provided for only mixed results.

In 1980, 52% of those surveyed were against Trident in a Marplan poll for the BBC's Panorama programme, and 59%

---

73 BBC News, Petition challenge to Trident legality, 11 December, 1999; Public Petitions Committee. 6th Meeting, Session 1, 14 December 1999 Available: www.scottish.parliament.uk/official_report/cttee/petit99-00/pumop1214.htm. (03/09/02)

against the system in a Marplan/Weekend World poll nearly a fortnight after.[74] In October 1982 a *Glasgow Herald* survey suggested that 55% of all surveyed were 'not in favour' of Britain acquiring Trident, though that same survey revealed that 56% of those polled believed Britain 'should have' a nuclear deterrent.[75] In 1984 a Gallup poll for CND suggested that 63% were against Trident, and by 1987 a System 3 poll commissioned by the SCND assumed that only 23% of Scots were in favour of the British continuing with Trident if the US and the Soviet Union agreed to withdraw medium-range missiles from Europe.[76]

In the post-Cold War era the picture becomes even less transparent. In 1991 a Gallup poll for Bradford University's Department of Peace Studies found that just previous to the 1987 general election, 67% of those polled would have been unhappy with only the two superpowers possessing nuclear weapons, while 55% supported a British deterrent. However, 69% of those polled were also opposed to the launching of missiles following a Soviet attack on the UK. Therefore, while a majority believed in retaining a British deterrent in this instance, an even greater proportion was not prepared to use these weapons.[77] Finally, previous to the May elections for the Scottish Parliament a Teletext poll carried out in April 1999 asked whether Scotland should have nuclear weapons, with 85% responding in the negative.[78]

While it appears that Trident was unpopular both in Scotland and throughout the rest of the UK, two surveys revealed that the British electorate believed it necessary to maintain a deterrent though one argued that the system should be maintained but not used. Firstly, if central government was not prepared to use its deterrent under the most extreme circumstances it ultimately undermined the very act of deterrence. Secondly, for those who suggested that Polaris should be retained while discarding the

---

74 S.R.T. Project 'Taking Out Moscow: Talking About Trident' (Edinburgh: St Andrew Press, 1991)
75 *Glasgow Herald* 18 October 1982
76 SRT 'Taking Out Moscow'
77 *Ibid.*
78 SCND Magazine, Available: www.cndscot.dial.pipex.com/magazine/nfs997c.htm, (02/07/02)

Trident option, it becomes obvious to central government that those surveyed appreciated the need for deterrence yet failed to grasp the repercussions of technical advancements with Russian capabilities at that time.

Disregarding the matter of usage, it was nonsensical to maintain a system that did not serve as an effective deterrent and central government believed that only Trident could provide the crucial advantage at that time. If we placed these results into a hypothetical situation where politicians actually executed policy under the guidance of public opinion, arguments against Trident were severely incapacitated. For many politicians these polls implied that it was nuclear war that was highly unpopular amongst civil society, and central government would argue that Trident was the reason this did not occur. Therefore, when considering the people of Scotland and their collective attitudes towards Trident one is forced to acknowledge that the poor evidence currently available provides no definitive conclusions.

The situation could therefore be described as follows: a minority either challenged or championed the deterrent but the overall reaction to Trident was characterised through disinterest and fluctuating degrees of ambivalence. With the Scottish media and sporadic opinion polls providing limited sources of information on defence issues it is likely that Trident's advocates take comfort from Scotland's confusion. Both the public and employees of shore establishments, largely unfamiliar with defence diversification and expanding technologies, were left in the dark.

VII
Things can only get better?

WHEN ATTEMPTING to predict the future of Coulport, Faslane or Britain's national deterrent one recognises that any hypothesis intended for classified defence projects are largely anecdotal unless extracted from somewhere close to the inner circle of power. Even then there is uncertainty until a formal decision is announced. Be that as it may, fun park attractions are prime examples of how theories can miss the mark.

At California's Disneyland, the 'House of the Future' was opened in 1957. It was an attraction that provided a glimpse of how we were supposed to be living in the future. You do not find traditional furniture styles or natural materials in the house. Everything is ultra-modern 1950s Americana and almost entirely synthetic. Enter the dining and family room and one finds a comfortable place where the family of the future will play, rest, and dine on stylish plastic furniture. Women are tickled by the chauvinist agenda which encourages them to utilise 'Atoms for the Living Kitchen', or microwave oven, as it is called today. We pass the bedrooms – one for the children of the future and one for the couple of the future – and conclude our tour in the sleek living room with its giant, non-operational, wall-mounted television screen. Some aspects of this attraction were definitely ahead of their time, while sexism and purely synthetic furnishings should be considered both dated and distasteful.

Yet when considering the future of Trident a similar degree of uncertainty becomes our predicament. Since the end of the Cold

War, nuclear issues have not been paramount in the minds of the public and commentators alike. Indeed, there is no present desire within central government, or Scottish Parliament, to revive nuclear weapons as the subject of public debate. If a decision on Trident is in the works, citizens and civilian employees might wish to have some degree of understanding about that which might impact their future.

The act of deterring aggression can be achieved through a number of methods of which conventional forces, diplomacy and economic sanctions are examples. However, at this stage of the game Trident is still considered the ultimate deterrent and it is in this context that British nuclear policy supports the concept of a minimum deterrent posture. Westminster has ratified the Comprehensive Test Ban Treaty and considers it a step towards the longer-term goal of complete multilateral disarmament as prescribed by the Nuclear Non Proliferation Treaty.[79]

Yet for Britain the national deterrent is largely a political tool and the rationale for Trident's retention is based on the system's reinforcement of the political process. Even during the frostier days of the Cold War, it could be argued that nuclear weapons encouraged the superpowers to pursue peaceful political solutions rather than direct engagement in the battlefield. Yet it must be realised that there are several key factors that will influence the government's decision, and the debate over Trident's replacement will not be about capability alone. The final verdict will be based on a combination of issues: current and anticipated strategic policies, Britain's international obligations (NATO), the benefits and cost of a replacement system, retention of the UK's supporting

---

79 The Comprehensive Nuclear Test Ban Treaty (CTBT) bans all nuclear explosions in all environments, for military or civilian purposes and was opened for signature in New York on 24 September 1996, when it was signed by 71 States, including five of the seven then nuclear-capable states. A major step towards the non-proliferation of nuclear weapons came with the signing of the Nuclear Non-proliferation Treaty (NPT) in 1968. Under the NPT, non-nuclear weapon States were prohibited from, *inter alia*, possessing, manufacturing or acquiring nuclear weapons or other nuclear explosive devices. All signatories were committed to the goal of nuclear disarmament.

infrastructure and whether it should remain a nuclear weapon state. One must recognise that disarmament requires more than just turning off the bomb when leaving the clubhouse.

It could be argued that the Trident system only has a political function in preventing a major war between nuclear weapon states. However, the Falklands War and Iraq have proven that nuclear weapons will not prevent war between proliferators and non-proliferators, and it seems that possession of a nuclear capability has a limited effect in stopping non-proliferators from trying to become proliferators. Furthermore, a sane person would assume that Trident certainly has no role against terrorism; it is difficult to present a realistic scenario where the SLBM might have a role today. Because NATO's 'Nuclear Forces in the New Security Environment' (NATO doctrine) argues that 'the circumstances in which their use might have to be contemplated is extremely remote', it ultimately questions the necessity for Trident.[80]

If Britain is to abandon its nuclear capability central government must feel absolutely secure in its decision. Currently, Iran and North Korea represent two stumbling blocks. If the UK disarmed itself, years of nuclear expertise would be lost and the supporting infrastructure would crumble. The cost of re-establishing the art and the article would be enormous. If Britain relinquished its capability it would also have profound implications for the Anglo-American relationship, giving credibility to those in America who never believed that Europe was committed to defending itself.

There are numerous considerations involved and any Westminster government only needs but one good reason to argue against Britain's surrendering of the minimum deterrent. It will probably retain a system so long as the Americans maintain cooperation for mutual defence purposes. and if interdependence should cease there is little question that London would seriously consider re-establishing its own fully independent deterrent.

For historical reasons it could be argued that at some point the UK will have to procure a replacement model. The submarine

---

80 NATO's Nuclear Forces in the New Security Environment can be found at http://www.nato.int/docu/handbook/2001/hb0206.htm

platform, recognised for its stealth, is still the most practical option. However, a changeover could involve a wider range of responsibilities for the current boats or a different class of submarine altogether.

Will Britain employ a single or hybrid platform delivery system? A replacement could involve a slightly modified or newly designed warhead to replace the current weapon. Is a Mark II version of the current warhead under consideration? Could the new system involve a mixture of conventional SLBMs with nuclear SLBMs? On the surface it seems that an escalation ladder is the most sensible option. Whatever course of action is chosen, it appears likely that the replacement will simply be an extension of the current system.

Nevertheless, one needs to recognise that civilian jobs in Scotland will have little if any influence on the decision-making process in London and that military planners are locked in a post-acute nuclear state of affairs where the situation is all but clear. But what should Scotland expect? If we assume that the upcoming replacement will be an extension of Trident, what could be some of the intermediate and long-term options? To highlight the uncertainty that inevitably hovers over the long-term usefulness of Scottish shore establishments, we shall first place our finger on the pulse of the US Department of Defence (DoD).

For the moment it appears that the DoD assumes the continued requirement for a sea-based strategic nuclear force. However, any defence project is subject to operational requirements; conditions that are ultimately dictated by the strategic setting. This atmosphere shifted after the Cold War and it will shift yet again. In January 2003 the US Trident force consisted of eight Ohio class Trident SSBNs based in Bangor, Washington, and eight Ohio class Trident SSBNs stationed in Kings Bay, Georgia, with another two Trident boats being converted into guided missile submarines, or Ship Submersible Guided Nuclear (SSGN) vessels.

It was the collapse of the former Soviet Union that allowed for this restructuring, and operational configurations envisioned for this new Trident SSGN class have only recently become clearer. The SSGNs will provide a multi-mission platform that can support both conventional land-attack and Special Operations Forces (SOF) in different force package options. In each of these, designated Trident missile launch tubes would be permanently modified for SOF diver lock-in/lock-out operations and attaching Advanced SEAL Delivery Systems (ASDS) or Dry Deck Shelters (DDS). The remaining missile-launch tubes will be modified to accept modules or canisters that would be loaded and unloaded just as Trident missiles are currently handled. Depending on the configuration, designated tubes would be loaded with either SOF stowage or a Tomahawk cruise missile module. According to the US Navy, there could be three basic force package options in total. Trident SSGN conversions are also thought to present a one-time, near-term opportunity to gain additional platforms of significant warfighting capability with low technical and schedule risk.[81]

Be that as it may, it has been reported that the Royal Navy is also preparing secret designs for a new generation of multi-purpose nuclear submarines to replace its own Trident fleet. According to the *Scotsman*, work on a replacement is further advanced than had previously been known, and the plan was to give ministers the chance to accelerate work on the new system so it could enter service around the year 2020.[82] However, if similar to the American SSGN design this does not sound like a near-term opportunity if deployment is not for another fifteen years! Though the replacement will probably be an extension of the current system (has this been mentioned yet?), there is a remote possibility that the Royal Navy is considering the conversion of one Vanguard class SSBN to an SSGN assignment with different intentions for the remaining three boats.

The US is currently engaged in a number of programmes designed to extend, improve or, in some cases, radically alter the

---

81 It appears that the initial steps to make the SSGN an official weapon system acquisition programme were taken when the US Congress appropriated $10 million in 2000 to commence design efforts on the ships. See: DoD releases selected acquisition reports for December 2002 period. See: http://www.globalsecurity.org/military/library/news/2003/04/mil-030412-dod01.htm

82 *The Scotsman* Secret plans for Trident replacement, 9 June 2004 See: http://news.scotsman.comtopics.cfm?tid=1034&id=653502004

capabilities of its Trident fleet. These programmes include enhancing the capabilities of the Trident SLBM, extending the lifespan of the system as a whole, beginning work on converting some Trident SSBNs to conventional use or improving the effectiveness of its warhead. Given the close cooperation between the UK and the US on all aspects of the Trident programme, Britain could choose to become involved in any, or all, of these programmes. Nevertheless, controversial new weapons may be on the more distant horizon. The Pentagon had intimated that a low-yield weapon, or 'mini-nuke', might be a new and safer option both for the military and civilians. There were lingering questions over the radioactive fallout, design, purpose and utility of this weapon but the Pentagon argued that this warhead would have been an excellent addition to its arsenal because it could have been used to incinerate underground bunkers containing chemical or biological weapons.[83]

If the MoD viewed this as a preferable option at some point, might low-yield warheads be one day situated within the Vanguard class? This cannot yet be determined. However, the UK government continues to assert that Trident will have a service life of roughly thirty years and if Aldermaston is considering the future of the current warhead there is reason to suspect that some type of new arrangement is already underway.[84]

If one relies on the Scottish media for information, one might suppose that Faslane could accommodate a Trident SSGN, a fully loaded Vanguard class 'mini-nuke' boat, another submarine armed with a combination of 'mini-nukes' and SLBMs and a fourth boat that acts as a standard boomer within the next few years. This degree of flexibility could be construed as a highly desirable option, but to claim this as a realistic alternative disregards Britain's current

83 In October 2005 the Bush administration abandoned research into a 'bunker-buster' warhead.

84 Though improbable, the *Scotsman* reported that the UK is on the verge of abandoning its long-range nuclear missiles in favour of cheaper 'mini-nukes' that could be used to strike rogue states. Again, it appears that an extension of the current system would be the most logical option. See at: http://news.scotsman.com/index.cfm?id=587652004

policy of maintaining one boat on patrol at all times. Even if the Royal Navy insisted on assembling this improbable mutation, what boat would be on patrol at what times and what would be the deterrent's specific mission?

Because a decision is not yet clear it is likely that we will discard this chapter and burn it in our fully synthetic fireplace. Such predictions are generally unreliable; that is the nature of the beast. However, we should widen our perspective and speculate as to what kind of future Coulport and Faslane might expect in the longer-term.

The Project for a New American Century, a neo-conservative think tank linked to the Defense-Intelligence establishment, the Republican Party and the powerful Council on Foreign Relations, has stated clearly that it seeks to revolutionise military affairs. This involves a number of advancements in various fields, including the concurrent weaponisation of space and the development of a new generation of nuclear weapons. The Missile Defense Agency not only pursues the successful completion of the controversial National Missile Defense programme, but also a range of airborne laser weapons with impressive striking capabilities.[85] Therefore, revolutionary developments in America may one day lead to drastic change in Scotland.

As technology evolves the nature of weapon systems will change as well, with a long-range forecasting study by the USAF identifying categories of smaller, lighter, agile, more lethal, and more affordable air-deployed precision weapon concepts achievable over the next ten to thirty years. This should be of interest to Scotland's civilian employees because it appears that the US is pursuing weapon systems that might one day replace the SSBN. The Pentagon's Force Application and Launch from Continental United

85 Missile defense systems being developed and tested are primarily based on hit-to-kill technology. Instead of an explosive warhead, hit-to-kill relies on a very high closing speed to collide with and destroy its target using only kinetic energy – the force of the collision to pulverize the target warhead. The Airborne Laser, or ABL, is a powerful laser aboard a modified Boeing 747 that destroys ballistic missiles by heating them until they fail structurally.

States (FALCON) programme, a reusable hypersonic cruise weapon system being developed by Northrop Grumman, has been described as a 'global reach' weapon that would allow the US to conduct effective, time-critical strike missions on a global basis without relying on overseas military bases.[86] Arguably, FALCON would allow the US to strike, either in support of conventional forces engaged in a war theatre or in punitive bombings directed against countries that do not comply with US economic and political diktats. Could the RAF be waiting for the day when the British could fund an interdependent hypersonic cruise weapon programme that returns responsibility for nuclear deterrence back to the skies? At the moment it sounds more like a weapon from one of Ian Fleming's Bond novels. However, before the SSBN there was the V-bomber and if the west of Scotland believes its relationship with the national deterrent is forever it might first learn to appreciate the fluidity of defence projections and the military's supporting industries.

Whether any of the mentioned observations or concepts proves to be correct is of little importance. What is important is that the west of Scotland is made aware of the impending situation. Coulport and Faslane will remain operational for some time but the brevity or longevity of their duration, purpose or function remains all too unclear. Therefore, it is necessary to break the spell that allows some to argue that the future of Strathclyde's shore establishments is impervious to change.

86 See: Northrop Grumman Takes Aim at Hypersonic Weapon Delivery System at http://www.irconnect.com/noc/press/pages/news_releases.mhtml?d=48827.

VIII
The big picture

IT APPEARS that the act of distracting or the condition of being distracted represents a problem for Western societies. In many instances methods of distraction are used to restrain information by flooding the media, or by inducing other people (or groups) to stop listening to a particular point of view. This can also be accomplished by drawing attentions elsewhere. In the age of the sound bite, it is far easier and more cost-effective to simply abandon an issue, rather than spending time or money on propaganda and spin.

Recently, several governments have discovered that the best way of dealing with the public is to accentuate the intentions of terrorists and 'rogue states'. Is it possible that the 'pacifist' label was somehow bestowed upon those who were opposed to the 2003 invasion of Iraq by those who had war on the top of their agenda? A similar situation applies to Trident as American and British authorities become ever more anxious over North Korean and Iranian nuclear capabilities. Yet distractions can lead to serious repercussions. In the short term authorities might need to consider some of the security issues relevant to shore establishments in Scotland. In the longer-term, Scotland needs to pay greater attention to the national deterrent because if Britain is going to continue down the path of deterrence its political leadership may eventually need to consider options that may not be favourable for the people of the west of Scotland.

Whether it be a conventional or nuclear conflict, the mechanics of going to war required communication, appropriate response and time to bring individual member states of NATO on to a war footing; the importance of Scotland's collaboration in these matters was nothing less than imperative. During the Cold War Scots generally understood that their homeland was riddled with immediate and major targets for Russian warheads, with the country committing itself to American, British and NATO security planning. Though Scotland's dependency on the British state still inspired questions of sovereignty, for many Russian targeting of Scottish territory ultimately validated the need for Trident. Because Scots were familiar with Polaris, and lodged firmly between the vice of economic depression and tensions with the USSR up until the early 1990s, the UK's adoption of the Trident system alarmed relatively few north of the Border.

Arguably, concerns with Polaris were allowed to dissipate sometime after the Americans occupied the Holy Loch, and before the installation of its British equivalent. There is no evidence of mass protest in Scotland associated with the Resolution class. Most in Scotland viewed the Polaris replacement as just that, with civil society generally unaware of the Vanguard's improved capabilities. That which was not seen nor heard sparked little interest. Trident, like Polaris, was a stealthy system tucked away on the Firth of Clyde, and with Scots preoccupied by their economic situation this allowed a majority to disregard its presence.

After the 1991 fall of communism in Eastern Europe, issues like 'first strike' and 'hard target' were considered irrelevant, and the Trident issue was no longer important to the Scottish electorate. Yet, in terms of security, it was assumed that Trident still served a purpose. That year hostilities in the Persian Gulf led to the invasion, and liberation, of Kuwait; the Americans left the Holy Loch in November and within weeks the lowering of the Soviet flag over the Kremlin symbolised a second Russian revolution as the Soviet Union fell. But with the passing of the communist superpower, it was Saddam Hussein's aggression towards its neighbour, including the use of chemical weapons against the Kurds in 1988, which tested the UK's defence posture. The system's continuity was seen as a reflection of the international community. Nevertheless, issues like devolution, the Poll Tax or the economy were of the highest priority in Scotland, and if Trident was decommissioned after the Cold War the British electorate might have considered this to be both a threat to national security and a monumental waste of the taxpayer's money. Britain's Trident would remain in Scotland. However, in economic terms the system served more as a life support system rather than a power station.

There is little question that Trident's contribution to the Scottish economy was over-rated by both central government and the Scottish electorate. World War I had brought with it a dependence on arms production, as global conflict provided for full employment in Scottish factories re-designed for war work. Yet despite the 'Red Clyde' phenomenon and the enhancement of workplace power, Scottish civil society at that time became a culture of survival. At the start of World War II the Scottish economy was not dissimilar from that which evolved during the previous years of conflict or post-war reconstruction. In the early 1940s the Secretary of State for Scotland, Tom Johnston, pressured the Churchill government into shifting significant war production back into Scotland. Again the country's economy was bolstered through arms production. However, after 1945 Scottish dependency on defence took on the nuclear dimension and continued amidst tensions with the Soviet Union, but this transition involved little for heavy industry. Though some conventional class boats were assembled in Scotland, submarine production brought little relief to the country as heavy industries were permitted to suffer a slow and painful death.

The drive to establish a means of support is not uniquely a Scottish characteristic in that there are countless examples of this throughout history. Yet in western society what is distinctly Scottish about this situation is the degree to which certain communities in Scotland relied on defence over the course of four decades. Even during the 'war on terror' these concerns will persist; in September 2004 there were intense fears that RAF Kinloss and RAF Lossiemouth faced permanent closure when it was announced that the MoD was reviewing its UK air force bases as part of its strategic defence review. For some this dependency was greater than for others. However, it appears that a majority of studies on the economic impact of military bases suggested that for every 100 jobs (service personnel and directly-employed civilians) created at a base there were between twenty and sixty jobs indirectly generated in the local economy. Yet a private export industry created twice as many jobs.[87] Nevertheless, at the end of the millennium the Scottish economy did not establish a reputation for adaptability.

Despite the discovery of the first 'Scottish' oil field in 1970, the brutal accountancy of market forces failed to either transform or prepare Scottish industry for oil production. In 1913 some 100,000 were employed in the metal manufacturing sector but by 1980 the steel industry in Scotland began to contract, with the Ravenscraig plant extinguished in 1992 due to the rationalisation programme of a privatised British steel. In 1913 there were 400 coal mines but by 1990 the Scottish miner was left with two. Between 1951-1991 the shipbuilding industry was purged of some 63,000 jobs.

In comparison, the national deterrent's effect on the Scottish economy was marginal. Scottish historians have written extensively on Scotland's transition from heavy industry to the electronics and financial services sectors in the 1990s, but, serving as testimony to the SSBNs impact on the Scottish economy, none have incorporated either Polaris or Trident into their economic analysis. Their silence over the national deterrent speaks volumes.

After the collapse of the Soviet Union it is certain that lasting economic interests were still key in Trident's ability to remain in Scotland undisturbed. Puerto Rico's experience with the US Navy is a useful comparison that might best demonstrate the veracity of this statement. For years there had been intense opposition to the American bombing range at Roosevelt Roads Naval Station on Vieques, and when the US Navy closed the facility in July 2003 the island lost the $250 million injected annually into its economy and the civilian jobs it provided. The opposition was successful, but whether residents of Vieques lived to appreciate their victory over their former American employer remains to be seen. When

87 M. Spaven *Fortress Scotland: A Guide to the Military Presence* (London: Pluto Press in association with Scottish CND, 1983)

questioned about the impact of the US Navy's decision to the Puerto Rican economy, Senator James M. Inhofe, Oklahoma Republican and member of the Senate Armed Services Committee, coldly stated, 'that's their problem, the time for them to be concerned about that was when they were kicking us off our range.'[88]

Shore establishments in the west of Scotland still serve a purpose but surrounding communities should consider the possibilities. Civilian employees opposed to nuclear weapons can be described as 'deterrence draftees' in the Trident project, and because the defence industry was vulnerable to a number of external influences Scottish workers were inevitably forced to accept that which was dealt to them. It must be remembered that Trident's relocation from Rosyth to Devonport cost £300 million more than originally projected and ultimately discredited Malcolm Rifkind's 1993 analysis.[89] Poor management was but one possibility that underlined the hazards of such dependency. A Scottish Parliament might keep this in mind.

If the public feels that figures recently released through the Department of Trade and Industry should hamper employment concerns with Trident, it might wish to consider the system's vulnerability to acts of terrorism before refilling its champagne glass.[90] Things have changed considerably since the end of the Cold War, and in 2005 the collective title 'Weapons of Mass Destruction' or WMD is now part of the political lexicon. Politicians from most parties in Scotland have been content to promote Britain as a nuclear power and major player on the European and transatlantic stage, originally as a bastion against the threat of communist domination but now in the so-called 'war on terror'.

---

88 *The Washington Times* 20 July 2003

89 Thirty seventh Report from the Committee of Public Account, Ministry of Defence: The construction of nuclear submarine facilities at Devonport, HC636 of Session 2002-2003

90 According to figures released through the Department of Trade and Industry, Scotland saw almost 5,000 new jobs created by foreign investors in 2004, and was ranked third out of the twelve main UK regions when it came to job creation by foreign firms. *The Scotsman*, Foreign direct investment, See: http://business.scotsman.com/economy.cfm?id=717042005

In the wake of September 11, 2001, the nuclear option has been intimately related to the fight against terrorists. As mentioned in the Bush administration's 2002 National Security Strategy (NSS), the pre-emptive 'defensive war' doctrine and the 'war on terrorism' against al-Qaeda was being used to constitute essential building blocks for the Pentagon's propaganda campaign in support of improvements with its nuclear technologies (the term 'mini-nukes' springs to mind).[91] In that context, the objective of America's NSS appears to be to prepare its public, and the world, for pre-emptive military action, or an act of self-defence, against two categories of enemies: 'rogue states' and 'Islamic terrorists', both of which are said to be in the process of obtaining WMD.

Yet with all this attention focused on US nuclear intentions and capabilities there is a distinct possibility that Islamic terrorists have considered Britain's nuclear deterrent to be an interesting opportunity. Why not strike first at that which may be used to strike you? If al-Qaeda has hit the Pentagon, and can put together another operation against the Trident system in Britain for less than a few thousand dollars, then why not launch an attack against the nuclear capabilities of a prominent member of the North Atlantic Alliance?

It may not be apparent to most Scots but stories about their country's anxieties with the national deterrent have been absorbed by the world's media. The efforts of CND in Scotland have not gone unnoticed, and a 1999 Greenock trial made international headlines when the Sheriff Court sided with the 'Trident three',[92] stating that the deployment and use of Trident 'could be construed

---

91 In October 2005 the Bush administration abandoned research into a nuclear 'bunker-buster' warhead. However, in a statement written for the president, 'The war against terrorists of global reach is a global enterprise of uncertain duration.' . . .America will act against such emerging threats before they are fully formed.' See Section V. 'Prevent Our Enemies from Threatening Us, Our Allies, and Our Friends with Weapons of Mass Destruction' of the National Security Strategy of the United States of America at: http://www.whitehouse.gov/nsc/nss5.html

92 Angela Zelter, Anne Moxley and Ulla Roder boarded a Ministry of Defence barge, a floating laboratory that monitored various signals that active Trident submarines emitted, and destroyed property worth an estimated £80,000. Zelter, A. *Trident on Trial: The Case for People's Disarmament* (Edinburgh: Luath Press, 2001)

as a threat and as such is an infringement of international and customary law'.

It appears that MoD property in Scotland is exposed to a number of difficult security issues. Yet if an attack on Coulport or Faslane shore establishments sounds far-fetched it may be helpful to review the past record of various Islamic groups.

Sometime in 1989 Osama bin Laden and Muhammad Atef founded 'al-Qaeda', the international terrorist group dedicated to opposing non-Islamic governments with force. One of the principal goals of this organisation was to drive American military forces out of the Saudi Arabian peninsula and Somalia through acts of violence. Furthermore, it has a command and control structure which included a *majlis al shura* (or consultation council) that coordinated terrorist operations; both Atef and bin Laden were on this council. The organisation also has terrorist groups operating under its direction, including: the al Jihad group based in Egypt, Al-Gama'a al-Islamiyya (the Islamic Group), formerly led by Sheik Omar Abdel Rahman, and other jihad groups abroad.[93] It is also believed that al-Qaeda forged alliances with the National Islamic Front in Sudan and with representatives of the Iranian government, and its associated terrorist group Hezbollah, for the purpose of working together against their perceived common enemies in the West. Though most of us are aware of al-Qaeda's intentions, the cowboys in Washington are still attempting to 'smoke 'em out of their holes' (or something to that effect).

It is instructive too, to consider how far al-Qaeda's operations reach. In February 1993 the World Trade Centre in New York City was badly damaged when a car bomb planted by Islamic terrorists exploded in an underground garage. The attack was carried out by followers of the Egyptian cleric Omar Abdel Rahman, who

93 Al-Jihad is an Egyptian Islamic extremist group that merged with Usama Bin Ladin's organization in 2001. Osama Bin Ladin's deputy, Ayman al-Zawahiri, was the former head of Al-Jihad. Active since the 1970s, their primary goal has been the overthrow of the Egyptian Government and the establishment of an Islamic state. The Islamic Group is responsible for many acts of violence, including the November 1997 Luxor massacre, in which 58 foreign tourists and four Egyptians were killed.

preached in the New York City area. In the August 1998 US Embassy bombings, hundreds of people were killed in simultaneous car bomb explosions at the US embassies in the East African capital cities of Dar es Salaam, Tanzania and Nairobi, Kenya. Everyone in Britain is familiar with the devastation caused some three years later in New York City and Washington. We can also recall the October 2002 Bali bombing in the town of Kuta where 202 people were killed and 209 injured. Jemaah Islamiyah, an Islamist group linked by news reports and Indonesian ministers to the al-Qaeda network, was immediately suspected of responsibility for the bombing. Finally, al-Qaeda itself claimed responsibility for the London bombings of 7 July 2005. With the IRA presumably abandoning its former methods of intimidation, the UK has been introduced to a bitter new flavour of terrorism. Given Britain's support for the US in Afghanistan, Iraq and elsewhere it would be irresponsible to underestimate the target value of Strathclyde's shore establishments for terrorist groups.

The Fleet Protection Group Royal Marines is responsible for defending the nuclear weapons in Faslane, at Coulport and in transit. The duty squadron provides a number of Reaction Forces and supports the static operations undertaken by the MoD Police. Without question, this is a top notch outfit that understands completely the Rules of Engagement, and is constantly introduced to new weapons, skills and responsibilities. However, while there is no question about their abilities, could it be that the positioning of these shore establishments and the easy access they support are subjects for concern in the current atmosphere?

Construction of the submarine base began in 1941 and left this facility hugging the eastern coast of the Gareloch in Faslane Bay. At that time the decision to build the Base in this location hinged on a number of geographical and operational factors. Being a facility that can be described as having an ellipsoidal perimeter, the A814 runs parallel to the Base and there is ample access to higher ground. Furthermore, one must remember that the Gareloch is not the English Channel; in some places the travel time for a motorised raft or jet ski from the Rosneath peninsula to the Base (and shiplift) can be estimated in seconds. While the entrance to Faslane now appears to be reinforced with a protective barricade, people who live in proximity to the Base are well aware that activists have entered the facility illegally (one protester even boarded a parked submarine), and that the entire establishment is left in full view of the outside world.[94] With the public's unrestricted access to Ardentinny, several of these conditions also apply to the Coulport facility.

There is the very real possibility that these shore establishments may be exposed to the dangers associated with various Package Type Improvised Explosive Devices (IEDs), Vehicle/Sea-Borne IEDs, Suicide Bomb IEDs and mortar attacks.[95] It must be realised that security concerns in the 1940s, 1960s and 1980s were very different from today's realities; what applies now were not the overarching concerns during World War II or the Cold War. In dealing with the cult of the premeditating martyr, it appears there may be a need for overhauling security procedures and extending the perimeter around these shore establishments.[96]

Within the contents of a MoD document dated 30 May 1956, it was stated that 'if local authorities were to be given the impression we had lost all interest in civil defence, then either the civil defence

---

94 There were urgent calls for a review of security at Faslane after two peace protesters infiltrated the base and daubed slogans on a Trident nuclear submarine. The incident was the most serious breach of security at a military base since the 11 September. There is little difficulty in determining if the shiplift is in, or has been, employed. See:http://thescotsman.scotsman.com/index.cfm?id=864662002

95 Package Types include time-delay triggered IEDs. IEDs that can be detonated by cordless phone from a car (allows for mobile firing platform and prevents tracing or triangulation). Vehicle borne IEDs (VBIEDs) are devices that use a vehicle as the package or container of the device. A suicide bomber brings a singularly difficult threat. The aim of the terrorist is not to commit suicide, but to kill or injure as many other soldiers civilians as possible.

96 In the 1980s physical security of shore establishments fell into five broad categories that included innocent trespass, criminal entry, espionage, protest groups, terrorist attack and sabotage. In those days protest groups were considered a relatively new threat to the physical security of military establishments, with control of demonstrations outside these establishments deemed a matter for the civil police and the Ministry naturally alert to 'deliberate trespass'. In 1984 the Defence Committee explained that [OVER]

services would begin to fade away or we should be faced with a demand for expensive measures to reinvigorate them'. This argument may apply to a contemporary setting. One must recognise that the chances of a Trident warhead detonating on MoD premises are vastly remote; to consider it feasible underlines the public's misunderstanding of the weapon's physics package or nuclear safing systems. However, there are several scenarios where immense damage of considerable expense can be envisaged for SSBNs that are moored on the Gareloch or cruising up the Clyde.

We are well aware that suicide boat-bombers blew a twelve by twelve metre hole in the side of the USS *Cole*, yet it appears that Trident SSBNS do not set out on patrol under heavily armed escort (a suggestion verified by residents who frequently witness the submarines' comings and goings). Should this tactic be applied to a vulnerable SSBN there is a remote (due to the integrity of the SSBN's pressure hull) possibility that hazardous materials associated with the SSBN could be released into the environment. In this scenario submarine safety training would almost certainly prevent the boat from drifting to the bottom of the Clyde, but a damaged boat might require towing.

If an attack caused a submarine to take in water the cost in submariner's lives could be high. There is little question that the political and psychological trauma caused to Scotland and the remainder of the UK could have serious implications for both Edinburgh and London. Filling security gaps at these facilities could involve significant expenditure, but, in the worst case scenario, an unexpected attack might lead to temporary or even permanent closure of these facilities (depending on the recommendations cited within a security assessment). Regardless of the damage a suicide mission may or may not produce, a single terrorist act could termin-

---

[CONTINUED] 'protest groups currently account for the great majority of unauthorised incursions' and that 'authorities have had to meet the nuisance caused by their activities by a variety of countermeasures'. Second Report from the Defence Committee, 'The Physical Security of Military Installations in the United Kingdom', HC 397-II of Session 1983-1984, p. vii. Also see: Second Report from the Defence Committee, 'The Physical Security of Military Installations in the United Kingdom', HC 397-I of Session 1983-1984

ate the 3,300 civilian jobs that both Coulport and Faslane support. But if Britain insists on operating its nuclear deterrent from Scotland how does the MoD go about defending these facilities without riding roughshod over the civil liberties of those associated with the disarmament movement?

In the wake of the events of September 11 and 7 July, it is certain that American and British governments are reviewing constantly their laws and security procedures, in order to try to reduce the risk of similar acts happening yet again, and/or to be seen to be doing something about terrorism. Hijackers associated with al-Qaeda have demonstrated with considerable success that a group of determined people can cause a massive amount of damage for a minimum investment. It was nineteen hijackers that forced the US to spend billions on its military campaigns in Afghanistan and, arguably, Iraq.

Clearly, preventing acts of terrorism from occurring is now a priority for the world's intelligence and security agencies. It is natural for laws to be revised in the light of this threat. But could activists in Scotland face new legal barriers when attempting to voice their opposition to Trident? What now should be done with activists who are caught disrupting warhead transports or crossing perimeter fences? Should their efforts to disrupt operations with the national deterrent be restricted to an even greater extent? And should anyone assume that al-Qaeda's sympathisers might blend in with organisations like SCND and use the disarmament movement to carry out their operations?

One could argue that anti-terrorism measures have done little to ensure Britain's safety or security from terrorist attack, but we cannot allow these measures to infringe upon the civil liberties of those living in the UK. New measures could serve to undermine centuries of democratic tradition, and become disastrously counter-productive. With the situation as such, it appears that there are some hard questions that have yet to be resolved. Yet should anyone in Scotland expect some form of action from the Scottish Parliament?

At both shore establishments there are security precautions in place that are not meant to be noticed, but it is important to note that there is no such thing as an impenetrable military facility.

The availability of the most modern technologies presents no guarantees. Therefore, if attacked the situation of Strathclyde's shore facilities would have to be seriously considered; attacks on underground trains and on buses are one thing but the national deterrent is another issue altogether. It should also be recognised that there is the distinct possibility that members of the Scottish Parliament have held discussions with various agencies on this issue, but because of the sensitivities Trident generates in Scotland it may have been withheld from the public arena.

If such talks were made public knowledge, they could lend credibility to the argument for decommissioning or relocation of these bases, and it just might present Labour in Scotland as being remarkably ineffective. If a Scottish Parliament was somehow encouraged to acknowledge various security difficulties with shore establishments, this ultimately places Scotland's Labour Party in a position where it may need to actually stand by its pledge in the Declaration of Faslane (a document which states that 'we mandate all our political representatives to rid Scotland of Trident').

Another argument is that by not addressing this issue publicly, thus allowing Trident's current situation to persist, it could serve as an unwelcome invitation to terrorist action.

Rectifying this problem could prove expensive. Amongst a number of improvements, this might require considerable expansion of the perimeters surrounding both Coulport and Faslane and the construction of new trunk roads. On 21 September 2005 (three months after the London bombings and four years after events in New York City) two residents of the Ardentinny area were asked if heavily armed escorts protected either the Coulport facility or SSBNs cruising down Loch Long. One response was, 'No, we don't see guns on these boats. It seems that a regular police boat appears from time to time. SBS – special boat squadron is sometimes there – small boats, but there are no visible guns.'

We are told repeatedly by the politicians that the war on terror could take years of effort, and that September 11 changed everything. Could it be that various governmental organisations, including the Scottish Parliament, have failed to heed their own warnings? With British soldiers currently serving in the Persian

Gulf, and a mass media serving its function by constantly seeking to entertain rather than inform, perhaps the public and its authorities are simply distracted.

Appendix

An interpretation of the CONCORDAT BETWEEN THE SCOTTISH MINISTERS AND THE SECRETARY OF STATE FOR DEFENCE

Firstly, the Scotland Act 1998, Schedule 5, Part 1, Paragraph 9, or Defence Reservation, clearly states that:

(1) The following are reserved matters –

a. the defence of the realm,

b. the naval, military or air forces of the Crown, including reserve forces,

c. visiting forces,

d. international headquarters and defence organisations,

e. trading with the enemy and enemy property.

Nevertheless, the Aim (3) of this concordat is to protect the 'general principles of the Scottish devolution settlement as they affect defence', and it provides a guide for 'arrangements for consultation, exchange of information, confidentiality and security, access to services, resolution of disputes, and review of relations'. Furthermore, its mission is to ensure that 'Scottish Ministers take into account the need for the unimpeded conduct of the defence of the UK, and the interests and responsibilities of the Secretary of State for Defence and the Armed Forces, when framing and implementing Scottish legislation or otherwise undertaking actions and functions within their competence'.

However, 'the interests, rights and responsibilities of the Scottish Ministers are recognised and taken into account by the Secretary of State for Defence in framing and implementing UK defence policy and activities likely to have an impact in Scotland'.

Should acts of terrorism on Strathclyde's shore establishments be a matter for concern, contingency planning (9) between Scottish Ministers and the Secretary of State for Defence allows for these parties to 'co-ordinate plans for contingencies such as civil emergencies, crises or war.'

Under Annex: Section 2 (A) Activities part (4) covering areas of co-operation between the MoD and the Scottish Executive, there is an illustrative list of areas where consultation between MOD and other authorities in Scotland occurs. The subject of 'counter-terrorism' is included with this.

Finally, should the Scottish Ministers seek an adequate awareness of developing defence technologies it appears that plans have been implemented for the Exchange of Information (16). According to the concordat, the Secretary of State for Defence and the Scottish Ministers should attempt to 'ensure that each is aware and takes account of the interests of the other. This may include policy papers, analysis and statistics, and other factual information. Should any doubts be raised about whether such information should be provided, the matter will be referred in the first instance to officials of Industrial Policy Division and the Regional Policy Unit for consideration.'

Greater research on this topic is necessary.

Bibliography

Arkin W.M. and Joshua Handler, Neptune Papers No. 3: Naval Accidents 1945-1988 (London: Greenpeace Institute for Policy Studies 1989)

Bell, I., 'Publishing, Journalism and Broadcasting' in Paul H. Scott (ed.) *Scotland: A Concise Cultural History* (Edinburgh: Mainstream Publishing 1993)

British Strategic Nuclear Force, The, Cmd. 8517 (London: Her Majesty's Stationary Office, 11 March 1982)

Central Unit for Procurement, The Trident Works Programme: A Case Study (CUP Guidance 1995)

Chilton, P., 'Nukespeak: nuclear language, culture and propaganda' in Crispin Aubrey (ed) *Nukespeak: the media and the bomb* (London: Comedia Publishing Group 1982)

Defence Committee, Ninth Report from the, The Progress of the Trident Programme, HC 237 of Session 1989-90

Defence Committee, Second Report from the: The Physical Security of Military Installations in the United Kingdom, HC 397-I of Session 1983-1984

Defence Committee, Second Report from the: The Physical Security of Military Installations in the United Kingdom, HC 397-II of Session 1983-1984

Defence Committee, Fifth Report from the: The Progress of the Trident Programme, HC 374 of Session 1988-1989

Defence Committee, Seventh Report from the: The Royal Dockyards, HC 829 of Session 1992-93

Dodd, T. and Mark Oakes, The Strategic Defence Review White Paper, International affairs and defence section, Cmd. Paper 3999 (1998)

Fife Regional Council, Rosyth: Memorandum of Evidence from Fife Regional Council to the House of Commons Select Committee of Defence (Fife Regional Council July 1993)

Gray, A. *Why Scots should rule Scotland : A carnaptious history of Britain from Roman times until now* (Edinburgh: Canongate Books 1997)

Harvie, C. *No Gods and Precious Few Heroes: Twentieth Century Scotland* (Edinburgh: EUP 1998)

Hawthorn, J. 'Some Thoughts on an Independent Scottish Defence Force': Occasional Paper No. 1 (The Scottish Centre for War Studies, University of Glasgow 1998)

Hutchinson, R. *Submarines, war beneath the waves: from 1776 to the present day* (London: Harper Collins 2001)

Johnson, R. *British Perspectives on the future of Nuclear Weapons* (Washington DC: Henry L. Stimson Centre 1998)

Leonard, Richard Speech 'The Economic Consequences of Trident' (March 1994)

MacDonald, I.O. *Faslane Facts and Feelings: A study of people's knowledge and attitudes* (Edinburgh: S.R.T. project, Church of Scotland, 1981)

McSmith, A. *John Smith: A Life 1938-1994* (London: Mandarin Press 1994)

Ministry of Defence and PSA, Proposed Development at the Clyde Submarine Base Faslane and Coulport: Environmental Impact Assessment (May 1984)

Ministry of Defence and PSA, Specification Report for Proposed Rosyth Nuclear Submarine Refitting Facility (April 1986)

Ministry of Defence, Notice of Proposed Development by the Ministry of Defence: Nuclear Submarine Refitting Facility at Rosyth Naval Base (Ref: DSD/NBD/750/16/86RD57, 1 May 1986)

Ministry of Defence, Thirty seventh Report from the Committee of Public Account: The construction of nuclear submarine facilities at Devonport, HC636 of Session 2002-2003

National Audit Office: Report by the Comptroller General and Auditor General, Management of the Trident Works Programme, HC 621 of Session 1993-94

PSA Defence Works (Navy) Docks and Main Works, HM Naval Base Rosyth Nuclear Submarine Refitting Facility: Supplement to Design and Landscape Strategy (August 1989)

S.R.T. Project 'Taking Out Moscow: Talking About Trident' (Edinburgh: St Andrew Press 1991)

Simpson, J. *The Independent Nuclear State: The United States, Britain and the Military Atom* (London: Macmillan Press 1986)

Spaven, M. *Fortress Scotland: A Guide to the Military Presence* (London: Pluto Press in association with Scottish CND, 1983)

Thatcher, M. *The Downing Street Years: 1979-1990* (New York: Harper Collins 1993)

Zelter, A. (ed.) *Trident on Trial: the case for people's disarmament* (Edinburgh: Luath Press 2001)

Chronology: The SSBN in Scotland 1960-2005

March 1960 Macmillan allows the US Navy's use of the Holy Loch as a forward-operating base for the Polaris deterrent.

March – May 1961 SCND assembles action against US presence at the Holy Loch.

October 1962 Cuban Missile Crisis.

April 1963 Polaris Sales Agreement concluded between US and UK governments. The Admiralty Polaris Committee also agrees that Rosyth should refit the new sea-based deterrent as well as other SSNs.

May 1963 Work to accommodate the UK's Polaris programme at Faslane begins, government places orders for Polaris boats.

September 1966 The first of the Resolution class Polaris boats, HMS *Resolution*, is launched.

June 1968 HMS *Resolution* undergoes first patrol from Faslane

September 1968 HMS *Repulse* commissioned.

November 1968 HMS *Renown* commissioned.

December 1969 HMS *Revenge* commissioned.

May 1972 Nixon and Brezhnev sign the first Strategic Arms Limitation Treaty.

August 1975 Ford and Brezhnev sign the Final Act of the Conference on Security and Cooperation in Europe between the US and USSR.

March 1976 Soviets deploy SS-20 missiles in Eastern Europe

January 1979 Callaghan and Carter reach agreement over Trident

March 1979 Scottish devolution referendum fails to pass with 40% approval.

July 1980 The GLCM is considered but decision to procure Trident I announced by Thatcher government.

November 1980 Manchester City Council joins NFLA, Glasgow and Edinburgh follow soon after.

March 1981 Labour splits, 'Gang of four' set to establish the SDP.

June 1981 First NOPD for Trident Works Programme released.

March 1982 Decision to procure Trident II announced by Thatcher government; Catholic Church condemns Trident; Argentineans invade the Falklands.

June 1982 Faslane Peace Camp established outside submarine base

September 1982 Missile fitting and servicing for Trident transferred to US facility.

January 1983 DS 19 established to counter CND influence.

May 1983 Strathclyde Regional Council's Coulport Inquiry released.

June 1983 Conservatives win general election with 144 seat majority.

May 1984 Dumbarton District Council joins NFLA; revised NOPD released.

March 1985 Secretary of State for Scotland, George Younger, approves Trident Works Programme.

May 1985 Asbestos discovered at Faslane; AESG releases first study on Trident.

May 1986 NOPD released for Rosyth Royal Dockyard.

June 1987 Conservatives win general election with 101 seat majority, but their Scottish seats plummet from twenty-one to ten.

March 1988 Liberal Democrat Party established; AESG release second study on Trident.

May 1989 Kinnock and Labour reject disarmament

October 1989 NFLA release Civil Defence Planning Assumptions in Central Scotland.

November 1989 Border between East and West Germany opened.

November 1990 Thatcher ejected after poll tax riots; John Major is PM.

December 1990 Major confirms dedication to project when he states states, 'We must have Trident'; Drell Commission warns of accidental explosion with Trident SLBM during flight.

November 1991 US Polaris fleet withdrawn from the Holy Loch.

December 1991 Soviet Union collapses.

April 1992 Conservatives win general election with twenty-one seat majority.

September 1992 Major government announces competition for refits

October 1992 HMS *Vanguard* arrives at Faslane.

June 1993 Rosyth stripped of Trident refits.

August 1993 Faslane opened for Trident; HMS *Vanguard* commissioned.

May 1994 HMS *Victorious* arrives at Faslane.

January 1995 HMS *Victorious* commissioned.

March 1996 HMS *Vigilant* arrives at Faslane.

July 1996 ICJ Advisory Opinion rules the threat or use of nuclear weapons violated humanitarian law.

December 1996 HMS *Vigilant* commissioned.

May 1997 Labour government returns under Blair with 177 seat majority.

September 1997 Scottish devolution referendum passed.

November 1997 Royal Assent given to Scotland Act 1997.

May 1998 Declaration of Faslane released.

July 1998 Strategic Defence Review confirms Labour's dedication to Trident.

May 1999 HMS *Vengeance* arrives at Faslane; Labour secures majority in Scottish Parliament. Scottish Parliament reconvened.

September 1999 HMS *Vengeance* commissioned.

October 1999 'Trident three' acquitted.

October, 2000 Three judges asked to rule that Britain's nuclear arsenal is legal. The lord advocate sought a judicial ruling in the hope that another court cannot make a similar decision.

March, 2001 Three Scottish High Court judges ruled Britain's Trident nuclear weapons are not illegal under international law

April 2001 Anti-trident demonstrators disrupted a sitting of the Scottish Parliament.

February, 2002 More than 100 people, including two members of the Scottish Parliament, arrested during a protest at Faslane.

November 2002 Two activists entered Devonport naval base, boarded HMS *Victorious* and rang the ships bell

September 2003 Public Accounts Committee revealed that Trident's relocation to Devonport cost £300 million more than expected

January 2004 Two activists breached high security at Faslane Base

May 2004 Faslane infiltrated by two early morning swimmers.

July 2005 Faslane protest – one man was arrested when he managed to enter the Base

September 2005 'What Dr Jamison is saying is hardly a blinding light on the road to Damascus. Neil Smith, MoD spokesperson

Index

V

W

Acknowledgements

I wish to thank Dr Irene Maver, Professor Evan Mawdsley, Professor John Simpson, Professor Christopher Harvie, Dr John Walker, Mr. Derek Armstrong and Mr Desmond Fik; each played pivotal roles in the creation of this publication.

I also owe a huge debt of gratitude to Pat Callaghan, David Corbett, Alex Falconer, Mhairie Hunter, Stewart Kemp, Iain Leitch and Carole McCallum.

I wish to thank John Powles, Robert Purdie, Dr. Bill Speirs and Rachel Squire; each has played a key role in advising me. Each has directed me towards sources of vital information and offered numerous perceptive insights that served to guide me.

Lastly but certainly not least, I wish to thank my wife, Tiny, my mother and even Donald 'piles of smiles' Jamison for the endless support they have provided over the years. *Aut Pax Aut Bellum!*

Other titles in the Discussions Series

Mending Scotland
– essays in economic regionalism
Christopher Harvie
ISBN: 1 902831 74 8
128 pages £5.99 paperback

The weird and wacky world of the Scottish economy – a country with the worst health in western Europe, the lowest recycling and one of the biggest black economies.

The contrast between the situation in Scotland and the (generally) sober world of European regionalism has led to Mending Scotland, an introduction to regional economics and its place in society and politics.

And how ordinary people, working together, can change it, and their community, for the better.

Deep Fried Hillman Imp
– Scotland's Transport
Christopher Harvie
ISBN: 1 902831 30 6
96 pages £4.99 paperback

This short book argues that transport in Scotland is deep-fried unhealthy, and is time-warped in the age of the car. Failure to address the issue has social, economic and ecological costs.

Yet there are solutions – and they are being applied elsewhere. Professor Harvie argues that developed rail, light rail and tramways are the progressive way for the future.

Other books from Argyll Publishing

Handful of Rogues
– Thomas Muir's Enemies of the People
Hector MacMillan
ISBN: 1 902831 89 6
288 pages £15.99 hardback

The 1790s were a turbulent decade in Europe. This age of revolution had repercussions in Scotland where the lawyer Thomas Muir of Huntershill led calls for democratic reform.

Muir's contribution, argues Hector MacMillan in this challenging reassessment, can not be understood divorced from a background of popular demand, both nationally and internationally, for a radical renegotiation of the contract between Governors and Governed.

Our Scots Noble Families
Tom Johnston
ISBN: 1 902831 01 2
208 pages £7.99 paperback

Irreverent, polemical and entertaining, the classic text on land-ownership in Scotland written by former Secretary of State for Scotland, Kirkintilloch born, Tom Johnston. Originally banned by John Menzies on first publication, first printings went on to sell over 120,000 copies.

This new edition has the original preface by Ramsay Macdonald and a comprehensive new introduction by Brian D. Osborne, that brings the land question and Tom Johnston's legacy up to date.

www.argyllpublishing.com